Helicopter Pilot Reports

Other books by *Aviation Week & Space Technology* Magazine

Military Aircraft Pilot Reports
Business & General Aviation Aircraft Pilot Reports
Commercial & Regional Transport Aircraft Pilot Reports

Helicopter Pilot Reports

Aviation Week & Space Technology Magazine

McGraw-Hill

New York San Francisco Washington, D.C. Auckland Bogotá
Caracas Lisbon London Madrid Mexico City Milan
Montreal New Delhi San Juan Singapore
Sydney Tokyo Toronto

McGraw-Hill

A Division of The ***McGraw-Hill*** *Companies*

Published by The McGraw-Hill Companies, Inc.

pbk 1 2 3 4 5 6 7 8 9 DOC/DOC 9 0 0 9 8 7 6 5

Library of Congress Cataloging-in-Publication Data

Helicopter pilot reports / by Aviation week & space technology magazine.

p. cm.

Includes index.

ISBN 0-07-003168-1

1. Helicopters—Evaluation. 2. Helicopters—Piloting.

I. Aviation week & space technology.

TL716.H446 1995

629.133'352—dc20 95-24991

CIP

Acquisitions editor: Shelly IC. Chevalier

Editorial team: Robert E. Ostrander, Executive Editor
Sally Anne Glover, Book Editor

Production team: Katherine G. Brown, Director
Susan E. Hansford, Coding
Janice Ridenour, Computer Artist
Toya B. Warner, Computer Artist
Lisa M. Mellott, Desktop Operator
Linda L. King, Proofreading
Jodi L. Tyler, Indexer

Design team: Jaclyn J. Boone, Designer
Katherine Lukaszewicz, Associate Designer

GEN1
0031681

Contents

Introduction

The pilot reports on helicopters come from the pages of *Aviation Week & Space Technology* and span some 20 years. The most recent report was written by David Green on the McDonnell Douglas Explorer. The first report contained in this compilation was on the Agusta A-109 in 1976 by Donald Fink, the current Editor-in-Chief of the magazine.

With the exception of the Explorer report, all of the other helicopter evaluations contained in this book were flown by *Aviation Week* pilot/editors. David Green, who flew the Explorer for the magazine, is a graduate of the U.S. Navy Test Pilot School at Patuxent River, Md. and primarily flew helicopters in operational service while in the U.S. Navy.

While Green is a graduate of a test pilot school, the members of the magazine's staff performing evaluation flights are not. The magazine's pilots flying the helicopters primarily come from either operational military or general aviation backgrounds.

Both Donald Fink and David Brown are private pilots and have flown helicopters and fixed-wing aircraft over the years. David Brown retired from the magazine after more than 30 years and most recently had served as Dallas and London bureau chiefs. Robert Ropelewski earned his wings in the Marine Corps and flew helicopters on combat missions in Vietnam. Nicholas Kernstock earned his wings in the U.S. Army. Both Ropelewski and Kernstock are no longer with the magazine.

The evaluation flights are always flown with experienced pilots, either a test pilot from a manufacturer or a service pilot highly qualified in the helicopter. Because of time constraints and insurance considerations, the magazine's pilots do not attend lengthy ground school or undergo a long check-out procedure required to fly a single-place aircraft, or as pilot-in-command.

Most evaluation flights are preceded by a discussion of the aircraft's systems and flight characteristics. With the newer helicopters, more time is being spent in looking at the avionics package in the aircraft to be flown. When available, a flight in a simulator is scheduled prior to the actual flight. The preflight briefing includes emergency procedures, the flight plan to be flown and an area check out. Following the flight, there is a debriefing of the flight, and often a discussion of the production and marketing plans for the helicopter.

The editor/pilot flying the helicopter is entirely responsible for writing the pilot report. The magazine does not have researchers to help gather information for the report. However, all pilot evaluations are read by the other pilots on the staff, and by the magazine's own technical and news editors prior to publishing. Reports are never submitted to the manufacturers or operators for checking or tweaking prior to publication.

Chapter 1

Military

Huey 2 targets foreign military market

David A. Brown/Ft.Worth
June 14, 1993

The U.S. Army and Army National Guard currently are studying the possibility of replacing or retiring a large portion of their utility helicopter fleet. Industry-funded programs are looking at new helicopters and modification of existing machines, both to fulfill the Army's perceived need and to capture a sizeable share of existing and future civil and international markets. This is the fourth Aviation Week & Space Technology *article on these developments. Previous articles which have been reported include the American Eurocopter Panther 800 (AW&ST Nov. 16, 1992, p. 46; the UNC/GE UH-1/T700 Ultra Huey (AW&ST Jan. 4, p. 44) and the Global Helicopter Technology Huey 800 (AW&ST Jan. 4, p. 48).*

UH-1HD Huey 2 has been designed primarily to meet the requirements of foreign military operators using earlier versions of the UH-1.

Huey 2 has retained the standard UH-1D 220-cu. ft. fuselage, although the extended nose section from the Model 212 can be added for more avionics equipment capacity.

Bell Helicopter Textron's UH-1HP Huey 2 is designed to meet the requirements of non-U.S. military operators, which the helicopter manufacturer believes will form the largest segment of the market for upgrades of existing medium lift rotary-wing aircraft. The modification, which Bell has designed to be done outside the factory, is expected to cost slightly more than $1 million per airframe, varying slightly with the avionics and operational instrumentation selected.

Bell has not ruled out the possibility that the U.S. Army will opt for upgrading some of its existing UH-1s under the pressure of budget restrictions likely to be imposed (AW&ST Apr. 5, p. 31). But the company believes that the only realistic possibility of sales in the U.S. at present will be to Army National Guard units now equipped with the UH-1D/H.

Accordingly, Bell and its partners, Textron Lycoming and Bogan Aerotech, have aimed the Huey 2 at two groups of nations—those whose military arms currently operate UH-1s, but which have expressed interest in improved performance, and those who do not now operate Hueys, but which have expressed interest in acquiring some through the Foreign Military Sales (FMS) program if performance can subsequently be improved.

"Interest in upgrading the Huey first came to us about four years ago from Latin American nations, which were operating the helicopter in hot and high conditions and needed better payload-to-altitude capability," Frank R. King, Jr., Bell's Latin America area manager, said. The operators interested in this type of upgrade had several requirements that Bell moved to meet.

"First, they wanted the manufacturer involved," King said, "because they were concerned about continuing support. Secondly, they wanted to keep and continue to use as much of the existing infrastructure as possible, and finally, they wanted an upgrade design which would allow them to do much of the upgrade work in their own shops and with their own technicians."

These considerations resulted in:

- A decision to retain the basic Lycoming T53 engine, upgraded to the T53-L-703 configuration. This provides the needed extra power for the desired hot/high operations and will allow the operators to continue using most of the spares and tooling already in their inventories. Initial upgrade of the existing engines must be done in a Lycoming-approved facility, but subsequent maintenance still can be done by the military operator as with earlier versions of the T53.
- Development of an upgrade kit, rather than a remanufacturing line at the factory, to convert the basic UH-1

Main and tail rotors and the entire power transmission system on the Huey 2 have been adapted from the Bell Model 212. Tail rotor has about 50% greater authority than similar rotors on other UH-1 versions.

Huey 2 is powered by an upgraded version of the Textron Lycoming T53, the T53-L-703.

D or H airframe to the upgraded version. Bell will send a team to the modification site to train the modification workers on any special techniques needed. No production tooling is required, and only one special conversion tool is needed. Modification work can be done at Bell-authorized service centers for nations that do not wish to do it in-house.

- Selection and qualification of a variety of flight and mission instruments and equipment that will allow each operator to tailor the Huey 2 to the customer nation's special requirements.

Bell currently is formulating proposals to be made to several Latin American nations, including Argentina, Brazil, Chile, Colombia and Venezuela, with pricing in 1993 or 1994 dollars. Other nations believed to be likely customers for the Huey 2 are Greece, Spain and Turkey among the NATO nations, and Australia, Korea, New Zealand, the Philippines and Taiwan around the Pacific Rim.

In developing the new version, Bell was reluctant to commit the investment capital needed to qualify all of the new equipment envisioned for the Huey 2, which led to the addition of a third partner to the program, Bogan Aerotech, Inc., of Houston and San Diego. Bogan assembled a team of specialist companies and financed the qualification of their products on the Huey 2.

These companies included Avionics Specialties, Inc., for turbine engine monitoring systems; Rockwell-Collins Flight Management Systems for advanced cockpit, mission management and navigation, and flight control systems; HeliConversions, Inc., for particle separator systems; Breeze-Eastern for rescue hoist and cargo hook systems; Westinghouse for day/night enhanced vision systems; Phaostron for night vision system-compatible instrument panel systems, and Smith Industries Aerospace for liquid crystal fuel quantity indicator systems.

The decision to retain the T53 engine, when other modified UH-1 versions and new production competitors have selected newer engine designs, was based first on the desire to retain as much of the existing UH-1 structure as possible and secondly, on the belief that the engine's fuel economics would not be a detriment to sales.

"We've had no indications so far that fuel specifics are a major concern to military organizations which now operate the UH-1," King said. "They tend to focus on mission capability, maintenance capability and compatibility with existing tooling."

Further, he noted, the Lycoming engine already is in the inventory of all Huey users, and the upgraded version will be able to make use of such items as manuals, spare parts now in stock and training already accomplished.

This tradeoff of advanced technology for familiarity is a key feature in the route that Bell is taking in marketing the Huey 2. It hinges on the T53-703's capability to provide the additional power to give the upgraded airframe the increased performance required to meet customers needs.

The T53-L-703 will, according to Lycoming specifications, provide an additional 250 shp. at sea level, 15C standard day conditions, and an extra 144 shp. at 6,000-ft. altitude and 35C ambient temperature. Military ratings, which can be used for 30 min., show an even greater increase. They climb from 1,400 shp. under sea level, standard-day conditions for the older T53-L-13B, to 1,800 shp. under similar conditions for the T53-L-703. Military power at 6,000 ft./35C conditions jumps from 922 shp. with the older version of the engine to 1,200 shp. with the newer version.

The Huey 2's transmission has been upgraded from a flat-rated 1,160 shp. in the older helicopters to a 1,290-shp. rating for five min. in the new and a maximum continuous rating of 1,134 shp. The entire power transmission system has been replaced with the Model 212, with more efficient gearboxes installed in the tail rotor drive system to increase tail rotor authority by approximately 50%.

This translates into some impressive performance increases including hover ceilings both in and out of ground effect and increased payload over a given range. The most spectacular increase is a 275% increase in hover ceiling in ground effect at 9,500 lb. gross weight on an ISA+20C day.

The older UH-1H could hover at approximately 3,000 ft. pressure altitude under these conditions, while the Huey 2 can hover in excess of 9,000 ft. On a standard day, the hover ceiling in ground effect is up nearly 40%—from approximately 11,000 ft. pressure altitude for the UH-1H, to more than 12,000 ft. for the Huey 2.

In practical terms, this means the Huey 2 can carry eight troops 125 naut. mi. and land them on top of an 8,000-ft. mountain at a temperature of 26C. The current UH-1H could fly the same mission, but could only carry one soldier in addition to the crew.

The Huey 2's maximum gross weight has been increased by 1,000 lb. to 10,500 lb., of which 642 lb. are extra payload. Maximum payload with full fuel now is 3,038 lb., up from the 2,396 lb. that the UH-1H can carry. Gross weight with an external load has increased slightly more, from 9,500 lb. in the present UH-1H to 11,200 lb. in the Huey 2.

I had an opportunity to fly the prototype Huey 2 from Bell's development and test center here recently with Bell Test pilot Dwayne

Tail rotor on the Huey 2 has a tractor layout, rather than the pusher design found on some earlier UH-1 versions.

Williams. It had recently returned from a demonstration tour in the Southeastern U.S. and was being readied for demonstrations in Europe scheduled before and after the Paris air show.

The prototype, bearing civil registration N16144, is equipped with what Williams termed a "generic cockpit," one equipped with many of the new instrumentation and avionics systems selected and qualified for use on the Huey 2, but not one intended for use on any specific mission. A Westinghouse forward-looking infrared (Flir) system was mounted between the two pilots' seats, with repeater screens on each side of the instrument panel. The cockpit was fitted with a Collins weather radar and a tactical avionics management suite (TAMS), also developed by Collins, which permitted easy control of all avionics on board from a single digital keyboard.

Outside air temperature was 18C, density altitude was 800 ft. and the wind was reported from the south at 10 kt. as we prepared for liftoff. We had 1,190 lb. of fuel on board, along with an added crewman, which brought our takeoff weight to approximately 7,890 lb.

Engine start system is simple, but different from earlier versions of the UH-1. A STARTFUEL switch is moved to the on position before

Instrument panel, a generic layout in the Huey 2 prototype, is designed to permit users to tailor cockpit instrumentation to their needs.

the starter is engaged, and it precisely meters fuel flow during the start sequence. It is turned off when the exhaust gas temperature reaches approximately 700–750C and the starter is released at 40% gas generator speed.

After the rotor was up to its nominal speed of 324 rpm, Williams lifted the helicopter off and demonstrated a series of 360-deg. pedal turns to the left while climbing to an altitude of about 350 ft. before making a south departure from the Bell heliport. The Model 212 tail rotor fitted to the Huey 2 has a wider chord than the blades on older UH-1 versions and is a tractor, rather than a pusher, design. Improved gear boxes in the power train combine with other factors to increase tail rotor capability by about 50% over the UH-1H.

We flew to a nearby training site, and I made two approaches to a hover to get used to the Huey 2's flight characteristics. In fact, at lower altitudes and slower speeds, the Huey 2 is virtually indistinguishable from older UH-1s.

"If it flies like a Huey, sounds like a Huey and feels like a Huey, it must be a Huey," Williams noted.

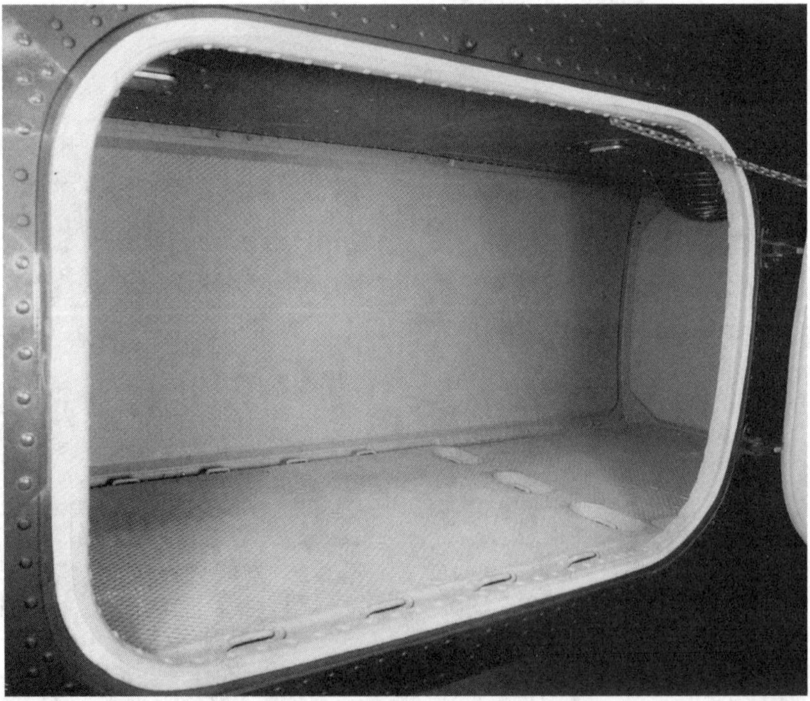

Added cargo capacity has been added to the Huey 2 by converting a six-foot-long section of the tail boom to hold cargo.

Westinghouse Flir unit, installed between the pilot seats, is an optional mission-oriented system to permit improved night vision.

I flew the approaches at 80 kt. on downwind and slowed to 60 kt. turning final and found the Huey 2 to be responsive and easy to maneuver. The only noticeable difference I could detect was the improved tail rotor capability at very slow speeds, which was initially a nuisance, since I tended to overcontrol. After a few minutes in hover, it became apparent that the improved lateral control capability it affords made possible precise positioning and heading in a hover. The wind had died away to light and variable breezes at this point, but Williams noted that it is possible to hover with the tail into the wind at wind velocities of up to 30 kt.

Williams demonstrated an autorotation, splitting the needles at 1,000 ft. and flying the helicopter at 65 kt. to the landing site before coming to a power-on hover, veering from right to left during the descent to demonstrate control authority with the rotor unpowered.

I flew the helicopter on a speed run over an approximate 20-mi. course to nearby Lake Arlington and return, getting the speed up to 128 kt. indicated air speed and found that, unlike some earlier UH-1 versions, the Huey 2 remains a docile and vibration-free helicopter as the speed increases. Never exceed speed has been increased on the

Main rotor system on the Huey 2 has been adapted from the Bell Model 212 and provides better lift capability and higher forward speeds.

Huey 2 to 130 kt., about 6 kt. higher than the present UH-1H. Williams noted that additional increases were possible as testing of the helicopter continues.

Vibration levels remained surprisingly low, again compared with earlier UH-1 versions, and Williams assured me that this was a feature of the rotor itself, which is not augmented by any sort of vibration damping system.

During this run we used the Westinghouse Flir system to pick out objects on the ground along our flight path and found that automobiles, pedestrians and other similarly sized objects could be identified.

We concluded our flight with a low-level, high-speed run through the trees that border the nearby Trinity River. This is a standard demonstration given with the Huey 2, meant to demonstrate the helicopter's combination of agility and speed.

Major changes incorporated in the Huey 2 from the earlier UH-1 series include the adoption of main and tail rotors, tailboom and horizontal stabilizer from the Model 212; a new main lift beam to permit the increased gross weights; a Kalflex drive shaft linking the engine and transmission; a baggage compartment in the tail boom; push-pull tubes instead of cables in the antitorque control system; an upgraded oil cooling system to handle the more powerful engine, and transmission and sturdier landing gear components to support the heavier gross weights allowed.

As far as the U.S. is concerned, Bell officials believe that the Army National Guard is the most likely to select the Huey 2. About 600 UH-1D/Hs are now in Guard units, and these constitute the second market that Bell is studying.

Conversely, Bell officials see only a limited civil market for the Huey 2. They note that U.S. operators already can avail themselves of independently developed supplemental type certificate (STC) modifications of the Bell 205 that give similar capability to that available in the Huey 2. In Japan, Fuji already has contracted with Bell for a Model 205 commercial upgrade for manufacture and use in that country.

"We believe there is an existing population of 800 to 1,100 UH-1s in foreign military service that could be converted to Huey 2s," King said.

BELL UH-1HP HUEY 2 SPECIFICATIONS

Maximum gross weight	10,500 lb.
Operating weight empty	5,440 lb.
Useful load	5,060 lb.
Fuel capacity, internal	1,435 lb.
Overall length, rotor turning	17.10 meters
Fuselage length	13.70 meters
Width, excluding rotor	2.60 meters
Height to top of rotor mast	3.60 meters
Main rotor diameter	14.63 meters
Tail rotor diameter	2.60 meters
Hover, out of ground effect, max. gross wgt., ISA conditions	5,600 ft.
35C ambient temp.	1,200 ft.
Max. speed, 10,500 lb. gross wgt., sea level, std. day	110 kt. IAS
7,500 lb. gross wgt.	130 kt. IAS
Max. range, std. fuel, sea level, std. day	240 naut. mi.
Maximum endurance	2 hr. 48 min.
Max. rate of climb, sea level, std. day, takeoff power	1,690 fpm.
Service ceiling	16,100 ft.
Cabin capacity, persons excluding pilots	11–13
Cabin capacity	220 cu. ft.

UNC-GE Ultra Huey vies for army utility role

David A. Brown/Ozark, Ala.
January 4, 1993

The U.S. Army and Army National Guard currently are studying the possibility of replacing or retiring a large portion of their utility helicopter fleet. Industry-funded programs are looking at new helicopters and at modification of existing machines, both to fulfill the Army's perceived need and to capture a sizable share of the future civil helicopter market. The following articles on the UH-l/T700 Ultra Huey and the UH-1/T800 Huey 800 are part of a series exploring these programs. The first appeared Nov 16, 1992, p. 46.

The effort to equip the U.S. Army and its National Guard and Reserve components with utility helicopters for the coming decades is centering on the selection of a new engine as much or more than the choice of a new airframe.

But, according to UNC Helicopter and General Electric Aircraft Engines, a proven engine with a modernized airframe and new avionics is the best combination possible. The two companies have teamed to develop the UH-1/T700 Ultra Huey—a highly modified and modernized UH-1 airframe, powered by a General Electric T700-700—as a competitor for the Army's utility helicopter in the late 1990s and on into the 21st century.

UNC/General Electric UH-1/T700 hovers during high-altitude, hot-day trials in Colorado. The helicopter is a basic UH-1H with a General Electric T700 turbine engine replacing the original Lycoming T53.

UH-1/T700 modifications include state-of-the-art cockpit instrumentation and K-Flex drive shaft to couple engine to transmission. Rotor, transmission, lift link, tail rotor and landing gear are from the Bell 212 design.

UNC Helicopter, based here, is a division of UNC, Inc. General Electric Aircraft Engines is headquartered in Lynn, Mass. Also on the team are Lucas Western, which developed a speed-reducing gearbox for the engine power transmission, and Bendix King and Canadian Marconi, which provide the avionics and instrumentation.

The Ultra Huey has completed hot-day/high-altitude tests in Colorado and now is performing flight demonstrations at Army and Army National Guard facilities around the country. UNC and GEAE have approached the need to upgrade the UH-1 to enable it to continue to serve with the Army and National Guard from two directions. UNC has worked to upgrade the airframe, avionics components and instrumentation. GEAE has taken the lead in developing and installing the new powerplant.

The team's goals for the upgraded helicopter included a 1,000-lb. increase in aircraft gross weight and growth in payload capability and range under certain conditions. Payload was to be increased by more than 1,600 lb. and range by 160 mi. under hot-day conditions.

UNC sought to incorporate the latest dynamic components from Bell's ongoing UH-1 developments in the Ultra Huey, while at the same time providing avionics and instrumentation in the cockpit that would go beyond any existing UH-1 configuration, according to Donald K. Flynn, UNC's project manager for the UH-1/T700.

Much of the airframe modifications have been taken directly from the Bell 212 version of the UH-1 helicopter. These include the rotor blades and rotor hub; tractor-type tail rotor, tail-rotor hub and tail-rotor control system; 45-deg. and 90-deg. gearboxes in the tail-rotor drive system; the main transmission assembly and lift link, and the landing gear. The 212 transmission is capable of accepting 1,290 shp.

The airframe also includes a UH-1N type tail boom, with strengthened tailboom attach points, and beefed-up fuselage and pylon structure. An ejector exhaust system has been devised and installed to help cool the engine bay.

Sling load capacity is demonstrated by the UH-1/T700 during hot and high tests in Colorado. The helicopter lifted 200 lb. more than predictions indicated. Total lift capacity is expected to exceed 10,000 lb. with an uprated engine.

The engine currently in the prototype Ultra Huey is a civil General Electric CT72, modified to military T700-700 specifications, and rated at approximately 1,325 shp. Already in production are more advanced versions of the engine, including the T700-701C/401C, which can produce about 1,890 shp. General Electric is developing versions of the T700 which will produce up to 2,500 shp., according to Samuel F. Powel, 4th, manager of T700 international programs.

GEAE also believes that the T700's current high rate of production—more than 500 units per year—will continue through the end of the decade. The engine is in use by the U.S. military in the Sikorsky H60 Black Hawk, the McDonnell Douglas AH-64 Apache and the Bell AH1W Super Cobra and in helicopters operated by the Australian, Italian and Japanese services.

The engine uses a hydromechanical fuel control system rather than the full-authority digital electronic control (FADEC) now employed on other engines, but GEAE officials believe this will not be a problem. The fuel control system does not require any rigging when installed and has demonstrated its capability to control the engine during a rapid "scramble start" in as little as 41 sec.

Of the engines competing to power the next utility helicopter for the Army, the T700 has the highest pressure ratio, delivers the most

Additional battery installation provides power for engine self-starts during extremes of temperature. The installation eliminates a need for external power.

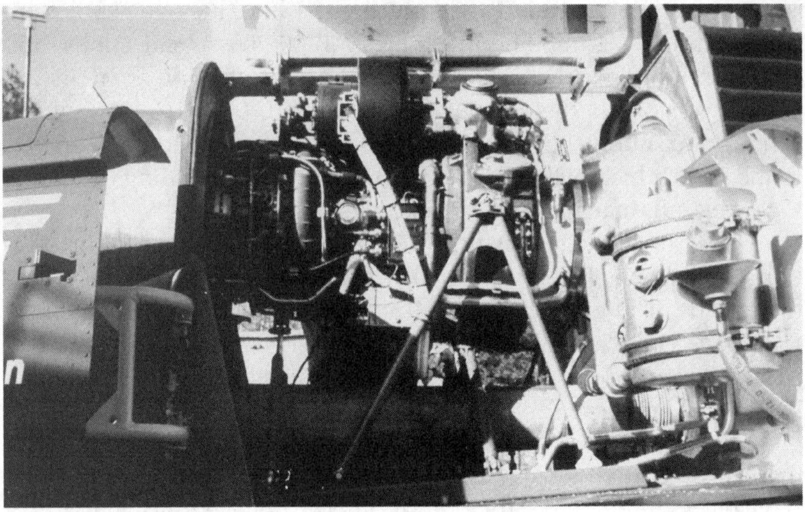

General Electric T700 engine, installed in the UNC-modified UH-1H, is a civil CT7-2 version modified to the T700-700 configuration.

shaft horsepower per pound of engine weight and will operate on less fuel, with specific fuel consumption about 25% better than the Lycoming T53-703 now in the UH-1 helicopter, according to GEAE figures.

The T700's fuel specifics are not as good as the Light Helicopter Turbine Engine Co. (LHTEC) T800, but this is more than offset by its ability to deliver more power under hot and high conditions, according to GEAE officials.

Furthermore, they said, the engine was proven during extensive operations during the Persian Gulf war in 1990–91, when it operated satisfactorily under extremely abusive conditions. Most of the problems encountered there were caused by first-stage compressor blade erosion or curling caused by ingesting gravel or other foreign objects, GEAE officials said. In about 90% of all problem cases, they added, the engine was repairable in the field.

Both Army and Navy statistics, quoted by GEAE, show the T700 has greater mean time between maintenance actions than other turbine engines in the two services' inventory, including turboshaft, turboprop and turbofan engines.

Added to the power system is a decreasing-speed gearbox. The unit in the prototype was developed by Lucas Western before the start of the Ultra Huey program and a new decreasing-speed unit, specifically tailored for the helicopter, now is under development.

The present unit reduces the engine's nominal 21,000-rpm speed to 6,600 rpm delivered into the main transmission gearbox. The engine is coupled to the transmission by a K-Flex drive shaft for lower vibration and more efficient power transfer.

The new avionics and instrumentation suite, which is being marketed as an optional extra, required a complete rewiring of the prototype airframe. The new cockpit incorporates a full-function flight management system, and the flight displays are night vision compatible. The vertical tape engine displays are the same as used in the UH-60 and AH-64, and the cockpit has two cathode ray tube displays for navigational and other data.

UNC also notes that the modified helicopter will have markedly lowered maintenance requirements from the earlier model UH-1s.

Much of this comes from the upgraded components adopted from the Bell 212. For example, the main rotor blades on the standard Bell UH-1H have an overhaul life of 2,500 hr., but this is increased to 4,000 hr. on both the 212 and the Ultra Huey. Main rotor mast lifetime jumps from 1,500 hr. on the Bell UH-1H to 15,000 hr. on the 212/Ultra Huey.

All other dynamic components, with the exception of the engine itself, will have overhaul lives of 5,000 hr. This includes the tail-rotor blades, the main transmission, and both the 45-deg. and 90-deg. gearboxes in the tail-rotor power train.

The T700 engine has no overhaul life—it is maintained on an on-condition basis, with maintenance work being done only when needed. The compressor and hot section are designed to operate at least 5,000 hr. between major maintenance, despite the fact that the power turbine uses uncooled blades. First- and second-stage turbine blades that drive the compressor are air-cooled, but are sand and dust tolerant.

UNC Helicopter and GEAE estimate that the cost of modifying a UH-1 to the Ultra Huey configuration would be between $1 million and $1.5 million, depending on the level of modification desired. They further estimate the cost of operating each unit for 300 hr. per year over a 15-year span would total about $3.58 million in present dollars.

I flew the Ultra Huey here shortly after the helicopter had completed its hot/high testing in Colorado and immediately before it departed for a tour of U.S. Army and National Guard facilities. I rode as a jump-seat passenger for a short flight from Blackwell Airport here to Dothan, Ala., Municipal Airport, about 25 mi. away. I then moved into the left seat, with UNC Helicopter Chief Test Pilot

Cockpit instrumentation and avionics layout, developed by Bendix King and Canadian Marconi, provides the pilot with two CRT displays. Note vertical tape display engine instrumentation.

Robert Whittaker in the right seat and Darrell O. Rowe, 2nd, GE manager of the UH1/T700 project, in the jump seat.

Start procedure is simple, from the pilot's point of view. The throttle is set to the "fly" position, the start button depressed and the pilot monitors the engine gas temperature to see that it does not exceed its 850C limit.

While simple in operation, the starter system on the Ultra Huey is more complicated in its design and layout. The T700 engine requires a high power level to achieve the 18:1 compression ratio needed to start it. The Ultra Huey has been fitted with two 26.5-v. batteries, mounted on the same racks as those needed for the T53. These batteries provide the power to make the helicopter independent of external power sources and allow rapid starts in extremes of temperature.

The starter automatically goes off line when the engine reaches 52–56% of its nominal compressor speed of 44,700 rpm. Engine output speed is 21,189 rpm into the speed-reducing gearbox, which lowers it further to 6,600 rpm going into the main transmission.

A revised version of the speed-reducing gearbox is now in development and is expected to cut the weight of the new engine installation when it is complete. The new box will be smaller, lighter and less complex and will not be power-limited.

The main transmission has a 5-min. limit of 1,290 shp. and a maximum continuous rating of 1,160 shp. The engine and rotor match-that is, the revolution indicator needles meet—at 100–104% of power output, with the rotor turning at a nominal 324 rpm.

Ambient temperature was 63F and the wind from the south at 3–5 mph when the Dothan tower cleared us to the taxiway adjacent to Runway 31. Takeoff weight was approximately 8,250 lb., or well below the maximum takeoff weight of 10,500 lb. We lifted off from the taxiway after receiving takeoff clearance and made a right turn to clear the Dothan traffic pattern, then proceeded to an experimental flight area approximately midway between Dothan and Ozark.

I flew the Huey Ultra in several gentle turns to get a feel for the controls before Whittaker demonstrated some of the helicopter's capabilities. The additional control power available through the tractor-type tail rotor was immediately evident and led me to overcontrol the helicopter for a time. Otherwise, the machine flew very much like the standard UH-1 of previous years. It was not possible to get an accurate fuel flow reading during this period, but the limited indications available showed that the T700 was probably using about 390–400 lb. per hr. or slightly less than the 400–440 lb./hr. anticipated in a typical 110-kt. cruise.

Whittaker demonstrated "rotor pulls"—rapid increases in collective pitch—to show that the fuel control system is capable of maintaining rotor speed even when additional torque is demanded from the engine within a 2–3 sec. period. While this is a capability well within the reach of FADEC-controlled powerplants, it is something that is unusual in engines like the T700, which rely on mechanical power scheduling.

An electrical control unit monitors the power turbine speed and feeds inputs into the hydromechanical unit to ensure that the engine is delivering maximum torque available, even when the power turbine is turning at less than 100% speed. Additionally, there is a load demand unit that monitors collective control position and schedules additional fuel flow through the hydromechanical controller.

In the prototype Ultra Huey, the torque output is controlled manually by the pilot, but in production versions a maximum torque safety scheduler is planned, which would automatically limit fuel flow through the hydromechanical engine control unit. In the event of a failure of the electrical unit, a lockout device allows the throttle on the collective to be used to control fuel flow on an emergency basis.

The Ultra Huey demonstrated its hot-day/high-altitude capability at tests near Alamosa, Colo., this past summer, where it lifted 8,700 lb. out of ground effect at 10,700 ft. altitude on a 59F day. The helicopter also lifted 9,720 lb. out of ground effect at 7,300 ft. on a 49F day.

After completing local demonstrations, we continued on to Blackwell Airport, where we shot several approaches, including one autorotational approach to hover. Rotor speed decay was not noticeable, but an excess of power in the tail rotor was, with some overcontrolling again evident.

The new cockpit layout in the Ultra Huey, although unfamiliar, was easy to read and an improvement over the standard UH-1 arrangement. The pilot's side, particularly, gains from the cathode ray tube displays and the vertical tape engine setting instruments.

UH-1/T700 ULTRA HUEY SPECIFICATIONS

Maximum gross weight	10,500 lb.
Operating weight empty	6,200 lb.
Overall length, rotor turning	17.62 meters (57.8 ft.)
Fuselage length	12.77 meters (41.89 ft.)
Width, excluding rotor	2.34 meters (7.68 ft.)
Height to top of rotor mast	3.60 meters (11.81 ft.)
Main rotor diameter	14.63 meters (48 ft.)
Tail rotor diameter	2.97 meters (9.74 ft.)
Hover out of ground effect, max. gross wt., ISA + 10	Exceeds 4,000 ft.
Max. speed, max. gross wt., sea level, std. day	120 kt. IAS
Max. range, std. fuel, sea level, std. day	430 naut. mi.
Cabin capacity, persons excluding pilots	11–13
Cabin capacity, cu. ft.	220

T800 engine boosts UH-1H's performance

David A. Brown/Ft. Worth
January 4, 1993

Global Helicopter Technology's Huey 800 meets objectives of greater fuel efficiency and safety.

The effort to equip the U.S. Army and its National Guard and Reserve components with utility helicopters for the coming decades is centering on the selection of a new engine as much or more than the choice of a new airframe. Global Helicopter Technology, Inc., of Ft. Worth believes that the Army and its reserves will not need a new helicopter during the next 20–30 years, just a new powerplant for existing Bell UH-1 airframes.

Global's Huey 800 is a basic UH-1H with a single Light Helicopter Turbine Engine Co. (LHTEC) T800 replacing the original Lycoming T53 powerplant. It currently is being demonstrated to Army and National Guard officials.

"It would be a crime to scrap the UH-1 helicopter airframes, many of which have only 5,000 or 6,000 hr. on them," Dan Pettus, president of the company, said. Studies by a number of helicopter manufacturers and airframe modification organizations have looked at a variety of new engines for the UH-1, after realizing that the basic airframe still has a lot of useful life.

Global Helicopter Technology Bell UH-1H with LHTEC T800 engine has a passive vibration damping system.

"The world situation has changed," Pettus said. "We need to seek missions that the UH-1, with a new engine in it, can do as well as, and more economically than, the Sikorsky [UH-60] Black Hawk."

Global's studies indicate there are about 3,000 UH-1s still in the Army and National Guard inventory and another 2,000 flying elsewhere in the world, in both civil and military colors. Of this total of about 5,000 UH-1s, as many as 20% are too old or heavily used to be worth modernizing, Pettus believes. Of the remaining 4,000 more or less, Global expects that between 50% and 80% will be modernized over a 15-year period.

"The engines now in the Huey are too expensive to operate, especially for civil operators who fly a lot, and the avionics are old," Pettus said. "But the airframe is proven and, in most cases, has a low number of flight hours."

Global is teamed with LHTEC, itself a joint venture of General Motors Allison Gas Turbine Div. and Garrett Engine Div. of AlliedSignal, in a joint venture to develop the Huey 800, a derivative of the Bell UH-1H with a single LHTEC T800 engine installed.

"We have designed a very low-cost helicopter conversion program that provides our main objectives of safety, economy and performance," Pettus said. After a two-year study, Global decided in October, 1991, that the performance and cost of the T800 warranted a fullscale development and certification program for a UH-1 derivative powered by the new engine.

"We realized, despite what others were saying, that the T800 was not too small to power the UH-1 efficiently," he added. "The engine gave the helicopter what it most needed—safety and economy." The economy stems from the T800's 40% lower specific fuel consumption than the engines now flying in the UH-1 and at least 20% advantage over any other engine being considered for retrofit in the Huey, Pettus claims. During the first 100 hr. of flight test, Pettus added, the T800 actually bettered the design goal of a 40% decrease in fuel specifics.

Converting a standard UH-1 to the Huey 800 configuration costs about $750,000 for the new engine and another $85,000 for modernized avionics, Pettus said. "Even if you add in the fuselage modifications needed to bring the airframe up to Bell 212 standards, you are still talking about less than $1 million for a fully modernized helicopter," he said. A Bell Model 212 transmission also can be installed, for additional cost. The 212 transmission is capable of handling more power than the standard UH-1 transmission.

Global has developed an engine retrofit kit that can be installed in 120 man-hours, and is offering the kit, instruction and training as part of a package. Installation of the Bell 212 transmission and modification of the fuselage to Bell 212 standards would take additional time. The training portion of the package will permit the purchaser to have two helicopters modified under supervision of Global engineers, at which point the purchaser's technicians would be qualified to perform the modification without assistance.

I flew the Huey 800 from the Global factory here at a time when it had flown approximately 90 hr., with another 17 hr. of running on a tie-down stand. Shortly after the flight, the Huey 800 logged about another 40 hr. during a tour of major Army and Army National Guard facilities in the southern and eastern U.S. More than 50 Army and National Guard pilots had an opportunity to fly the helicopter during this tour.

Ron Williamson, developmental test pilot on the Huey T800 project, was in the right seat, while I flew from the left. The surface temperature was near 90F and the wind virtually calm, blowing from the northwest at 1–3 kt. The helicopter weighed approximately 7,600 lb. at takeoff, about 1,900 lb. below its maximum gross weight of 9,500 lb. Maximum payload capability is 3,800 lb.

Williamson demonstrated the rapid start capability of the LHTEC T800, which differs considerably from the engines in older UH-1s. The engine start procedure consists simply of moving the power lever to the 100% power setting and depressing the start button on the cyclic control. The engine reached full operating speed of 6,600 rpm. in about 1 min.

The T800's dual-channel full-authority digital engine control (FADEC) monitors the measured gas temperature (MGT) to limit engine temperature, and an automatic torque limiter comes on line with software that restricts engine torque output to 45–46% during the start procedure.

The system has provision for a manual engine start, but the automated "scramble start" is normally used because it is simple and imposes safe limits on the engine during the start procedure. Using the FADEC to control the engine, the T800 can be started automatically over wide temperature and altitude ranges and can accelerate from a noload condition to takeoff power in about 3 sec. Engine speed will be controlled by the FADEC to within 0.5%.

Williamson lifted the UH-1H off the Global Helicopter Technology helipad, which is located immediately adjacent to several buildings, and established us on a heading toward Ft. Worth's Alliance Airport before turning control over to me. It was immediately apparent that this Huey, a veteran of Operation Desert Storm and now wearing civil registration N800NT, does not fly like the standard Huey I recalled from years past.

We cruised at 115 kt. indicated airspeed en route to Alliance and felt little of the vibration that usually is associated with the Huey at that speed. The familiar rotor slap remained, but vibration has been greatly reduced.

Williamson noted that the Huey 800 has been fitted with a passive vibration reduction system that is intended to limit the helicopter's two-per-rev vibration without sacrificing performance. The vibration reduction system uses a bobweight concept and requires no electrical or other power from the helicopter's other systems.

During the flight to Alliance airport, we trimmed the rotor speed to 320 rpm, or 99% of the standard 324 rpm, and Williamson demonstrated sharp collective pitch increases—"rotor pulls"—to show the engine control's capability to prevent rotor speed droop during periods of high power demand.

Williamson pulled 95% of available torque from cruise configuration, and the Huey 800 showed a rotor speed droop to approximately 314 rpm, or about 97% of standard. This maximum 2% transient rotor

Tennessee Army National Guard facility in Smyrna was among the Army, National Guard and Reserve units visited by the Huey 800 during a demonstration tour.

droop during rapid collective movements can be maintained during virtually all flight conditions, Williamson noted.

The FADEC system on the T800 engine senses the demand for additional power by monitoring both the speeds of the engine gas turbine and the rotor speed, and the collective position change rate. It commands engine power output to increase at a rate necessary to hold rotor speed droop to a minimum. We monitored this capability during both a power-on landing and an autorotation from altitude and in a rapid lift-off from a hover.

During the initial approach to Alliance, rotor speed was maintained at 314 rpm and airspeed at about 80 kt. until we were ready to flare into a hover. Speed then was gradually decreased as the nose of the Huey 800 came up to enter the hover.

I then took off and made two more approaches and landings to a hover at the field, and Williamson demonstrated the Huey 800's autorotational capability. We entered autorotation while turning onto a short base leg at an altitude of 1,400 ft. with the collective fully down and the power off, causing the needles on the engine/rotor speed display to split. We held 60 kt. indicated airspeed during the descent to one of the taxiways at Alliance airport until the Huey 800 reached 200 ft. agl.

At this time, the throttle was rolled to bring on power and the collective pulled to slow the descent rate and bring the helicopter into a hover. The FADEC compares gas turbine speed and rotor speed as the collective comes up. When the needles are about to realign—within about 2% of each other—it cuts back the engine acceleration, regardless of what the throttle is demanding, to ensure a smooth engine/rotor engagement.

From this hover, I flew the Huey 800 through a 360-deg. pedal turn in each direction to confirm that it has adequate control power with the T800 engine. Control was similar to the earlier versions of the UH-1, with tail rotor control power limits probably just about reached. The tail rotor, which has not been changed for this version, has the same limitations as the basic UH-1H, most of which are noted only at high altitude.

During these turns, Williamson pointed out the Huey 800's ability to overcome static rotor droop. Despite the need for power to be diverted to the tail rotor, main rotor speed did not vary during the turn and the engine did not "search" for the correct power setting. Williamson also pulled power via the collective from a fully down position while sitting on the ground to about 95% torque within about 2 sec. and got only slight loss of rotor speed.

The T800 engine delivers a maximum of 1,300 shp. in this installation, and the transmission in the Bell UH-LH is limited to 1,160 shp. The engine can be controlled manually if necessary by completely cutting the electronic fuel control system out of the control pattern via a switch on the collective control. The valve controlling fuel flow then can be operated by a beeper switch to adjust engine speed.

Global says the T800 engine already is exhibiting up to a 40% improvement in fuel specifics from engines in standard UH-1s. The T800 burns between 360 and 385 lb. of fuel per hour. The engine installation also has simplified maintenance procedures, since all field-replaceable components on the powerplant can be removed in less than 30 min. without using torque wrenches. No safety wire is needed when the engine is installed.

Global plans to obtain FAA restricted category certification on the Huey 800 and possibly military qualification as well during the next 100 hr. of test flying, which is now underway. Full civil certification then will be acquired, according to present plans, during the next 50-100 hr. of flight test and will be followed by life cycle testing.

GLOBAL HELICOPTER TECHNOLOGY HUEY 800 SPECIFICATIONS

Maximum gross weight	9,500 lb.
Operating weight empty	6,200 lb.
Overall length, rotor turning	17.62 meters (259.12 ft.)
Fuselage length	12.77 meters (41.89 ft.)
Width, excluding rotor	2.34 meters (7.68 ft.)
Height to top of rotor mast	3.60 meters (11.8 ft.)
Main rotor diameter	14.63 meters (48 ft.)
Tail rotor diameter	2.59 meters (8.5 ft.)
Hover out of ground effect, max. gross wt., ISA + 10	2,500 ft.
Max. speed, max. gross wt., sea level, std. day	114 kt. IAS
Max. range, std. fuel, sea level, std. day	360 naut. mi.
Cabin capacity, persons excluding pilots	11–13
Cabin capacity, cu. ft.	220

Notar reduces pilot workload, improves response in OH-6A

Nicholas C. Kernstock/Mesa, Ariz.
February 13, 1989

The McDonnell Douglas Notar concept offers reduced pilot workload and smoother control response than conventional, tail-rotor-equipped helicopters. These attributes were evident during a flight in the no-tail-rotor demonstrator helicopter here recently.

The Notar antitorque system has also nearly doubled the rearward and sideward flight envelope of the basic Army/McDonnell Douglas OH-6A aircraft. Additional improvements expected from production versions are lower life cycle and maintenance costs and reduced noise levels. McDonnell Douglas plans to use the system on two new commercial helicopters as well as the Bell Helicopter Textron/McDonnell Douglas LHX aircraft.

This *Aviation Week & Space Technology* pilot flew the Notar on a 45-min. demonstration flight at the McDonnell Douglas plant in Mesa, Ariz. Bert Rhine, McDonnell Douglas Helicopter experimental test pilot, accompanied me.

U.S. Army/McDonnell Douglas Helicopter Co. Notar-equipped OH-6A hovers backward into trees to demonstrate its safety in confined areas. The face of the variable pitch fan and the inlet screen can be seen at the base of the tailboom.

The aircraft's configuration forward of the tailboom is similar to that of the original OH-6A. The four-blade, articulated main rotor and transmission have been retained, as have the flight controls.

The cockpit is similar to the older military version except for the addition of six control position indicators mounted on the left of the instrument panel and the hydraulic pump switches and warning light at the lower right side of the panel. The OH-6A has unboosted controls, and the hydraulic pump was added to drive the variable-pitch fan in the tailboom. The back seat was removed to allow the installation of test instrumentation.

The airframe is the second OH-6A produced in 1963 by Hughes Aircraft and has been modified to include the installation of the pointed MD 500-series nose, an Allison 250-C20 engine and the composite Notar tailboom.

The Notar system uses a circulation control tailboom, a direct jet thruster and a variable pitch fan to counteract torque and provide directional control at low speeds and at hover. A brief inspection of the aircraft indicated one of the key advantages of the system—simplicity.

The thruster is a variable area nozzle that uses a rotating outer sleeve which is connected to the foot pedals to change the direction and amount of thrust. At full left pedal, the left thruster is completely open. Full light pedal rotates the sleeve 180 deg. to the right, opening the right thruster. At a hover, the pedals are nearly neutral and the thruster is almost closed.

Notar concept demonstrator accelerates prior to initiating a quick-stop. The jet thruster outlet, visible at the rear of the tail, is in the closed position.

A conventional cable and bellcrank transmit pedal control movements to the thruster. A quick check was sufficient to verify the freedom of travel of the outer sleeve and general condition of the horizontal and vertical stabilizers.

The circulation tailboom provides most of the antitorque force needed at a hover. Longitudinal slots vent two jets of air down the right side of the tailboom. The main rotor downwash is entrained by these jets and follows the contour of the tailboom. The effect is similar to that of a wing, creating lift toward the right and counteracting main-rotor torque.

The fan, installed in the base of the tailboom, is protected by a wire screen and is easily inspected. The production version will have smaller openings and should produce less noise. The pitch of the fan, controlled by a mechanical link with the foot pedal thruster control, is increased as the opening is enlarged to maintain constant pressure. High-speed rotating components, bearings, shafts and supports have been nearly eliminated.

I took the left seat for the flight, while Bert Rhine occupied the right seat. Pre-start checks and engine lightoff were simple and almost identical to the OH-6A baseline aircraft. With the rotor at full rpm, the lack of tail rotor noise and vibration was immediately noticeable.

High yaw rate

The aircraft's weight was about 2,500 lb. That included an empty weight of 1,900 lb., 200 lb. of fuel and a combined crew weight of about 400 lb. Normal maximum gross weight is 2,550 lb. Conditions at the McDonnell Douglas heliport were 58F and winds from the west at about 10 kt. Humidity was below 50% and the field elevation is 1,370 ft.

Rhine first flew some of the maneuvers to demonstrate the unique handling characteristics of the Notar helicopter. One impressive feature was the high yaw rate possible with the Notar. Rates in excess of 100 deg. per sec. were demonstrated with no loss of attitude control.

Also impressive was the availability of full left or right pedal travel. Use of full pedal on conventional helicopters can overtorque the engine, cause structural damage or stall the tail rotor, resulting in loss of antitorque control.

Hovering over a point on the heliport, we executed several 360-deg. full-travel pedal turns to the right and left. Maintaining position over the spot required only slight cyclic corrections to compensate for wind drift. There was no noticeable pitching of the aircraft as the tail turned into the wind.

Agility of the baseline OH-6A helicopter is retained in the Notar. The aircraft is shown making a high-angle bank turn. Notar reduces the need for control inputs in this maneuver.

With a tail wind, the turbulence from an exposed tail rotor would be blown into the main rotor downwash, pitching the nose down and spoiling lift. In the space of a few seconds the pilot would have to add power and collective pitch, compensate for the torque by adding left pedal and maintain level pitch attitude with aft cyclic. None of this was needed in the Notar.

Power demand for pedal turns was modest, drawing only 5–7% more torque for left turns (against engine torque) than for right turns. I initially had a tendency to overcontrol the fast pedal turns because the Notar required less control input than the standard OH-6A. After starting the turn, I applied opposite pedal to stop and then neutralized the controls. No additional corrections were necessary to stabilize the aircraft in a hover.

The aircraft did require a large amount of opposite pedal to stop the turn. This is because the tailboom is considerably heavier than the standard OH-6A tail. The glass fiber and composite Notar weighs about 300 lb. versus less than 100 lb. for the standard tailboom. Balancing the heavy tail is 250 lb. of lead installed in the nose of the aircraft. Rhine said the production aircraft tail sections will weigh only about 100 lb. This should eliminate the high momentum characteristic of the concept demonstrator aircraft.

Light control forces

Because the thruster sleeve is not springloaded and no other aerodynamic forces are changing the opening of the jet, control forces were light and did not require constant pressure. I only had to set the desired amount of pedal control and could remove my feet from the pedals.

Notar holds an out-of-ground-effect hover over the desert near the McDonnell Douglas plant in Mesa, Ariz. This position can be held with minimal use of pedals because of the tail boom design.

To demonstrate the antitorque effect of the circulation tailboom, I set the pedal indicator at 40% while the aircraft was on the ground and facing into the 10 kt. wind. With my feet off the pedals and on the floor, I brought the aircraft to a 3-ft. hover. No additional pedal input was needed to maintain heading.

I then pulled the aircraft to a 20-ft., out-of-ground-effect hover, still without making any changes to the pedal setting. The aircraft remained within 3 deg. of the original heading. Reducing the collective pitch control, I resumed a 3 ft. stabilized hover and then landed. The nose of the aircraft was 5 deg. left of heading. This maneuver simulates a masking tactic required in an antitank observation or attack situation.

Rearward flight of up to 50 kt. and sideward flight of 40 kt. was smooth and stable. Starting at a 10-ft. hover, we applied aft cyclic and increased power to approximately 95% torque. The aircraft went through the typical shudder of translational lift as the main rotor began flying at about 25 kt. Heading was maintained with only slight pedal movements. We stopped the aircraft with forward cyclic in about 100 ft.

I lined up the aircraft across the runway into a quartering right headwind. To initiate left sideward flight, I moved the cyclic to the left and added about 60% left pedal. Readjustment of the pedals was not required until I brought the aircraft back to a stationary hover.

Flight to the right was against the wind and required nearly full right pedal to maintain the proper heading. Speeds of 40 kt. are possible with Notar versus 20 kt. in the tail rotor-equipped helicopter.

Hovering autorotations were docile and did not require the same immediate application of right pedal as the baseline OH-6A. With the pedal set at 40% travel and the nose into the wind, we rolled the throttle to the idle position. Allowing the aircraft to settle a moment before applying approximately 3% right pedal, we cushioned the aircraft onto the ground.

After touchdown, only 10% additional pedal was used to maintain alignment. Notar allows a more forgiving and less abrupt hovering autorotational procedure. Straight-in autorotations have not been done in the Notar aircraft because of the tail-low configuration, Rhine said.

I executed several quick-stops to evaluate directional control when the tail is exposed to high airspeeds and a nose-high attitude. I accelerated from a hover and flew down the runway at an altitude of 10 ft. At 70 kt. indicated airspeed, I initiated the quick-stop by reducing collective pitch and engine torque to about 30%. Applying aft cyclic, I brought the nose of the aircraft up to 60-deg. pitch. Using right pedal to maintain heading, I allowed the aircraft to rapidly decelerate. I then pulled the collective pitch back up to prevent the aircraft from descending and leveled the nose as the aircraft stopped.

This maneuver requires a high degree of coordination and is used frequently in nap-of-the-earth flight. After the second quick-stop, I was able to hold heading and altitude without large control movements. Operation of the antitorque pedals were smooth and effective throughout the maneuver.

Easier pirouette

Several other high-pilot-workload maneuvers showed the improved handling of the Notar. The pirouette requires the pilot to hover in a circle around a fixed point with the nose of the aircraft pointed toward the center of the circle. He must maintain a constant height, rate of turn and distance from the point. This usually calls for frequent tail rotor adjustments but because the Notar is relatively unaffected by wind, these corrections are not needed.

The 360-deg. turns while hovering down the runway also required no power or antitorque adjustments to maintain a steady track, rate of turn and altitude. In conventional helicopters this maneuver demands constant movements of all the flight controls.

McDonnell Douglas will incorporate Notar in its five-place, single-turbine MD 520N and eight-place, twin-turbine MDX helicopters (AW&ST Jan. 23, p. 16).

Mast-mounted sight, new rotor boost effectiveness of OH-58D

Robert R. Ropelewski/Ft. Worth
September 8, 1986

Increased power, new four-blade rotor system and mast-mounted sight incorporated in the Army/Bell OH-58D interim scout helicopter provide substantial improvements in performance and agility over earlier Army scout helicopters while reducing the aircraft's exposure and vulnerability in combat.

Attitude, heading and other basic flight information are displayed on the copilot/gunner's multifunction display in the OH-58D. This is one of several formats that can be selected to be shown on the display.

Bell Helicopter Textron has completed delivery of the first production lot of 16 OH-58Ds to the Army and has begun delivering a second lot of 44 aircraft as part of an anticipated procurement of 578 aircraft. The OH-58D was earlier referred to as the Army helicopter improvement program (AHIP), and involves major modifications to existing Army/Bell OH-58A Aeroscouts rather than procurement of entirely new aircraft.

Additional equipment

Modification and improvements include addition of a mast-mounted sight with television and infrared sensors and a laser designator/rangefinder; larger engine and four-blade rotor with improved hot-day and nap-of-the-earth performance, and space, weight and power provisions for incorporation of General Dynamics Stinger air-to-air missiles. With these improvements, the OH-58D is expected to provide a day/night/adverse weather command and control, surveillance and target acquisition capability, and a capability to designate targets with its laser for Army and Air Force laser-homing munition.

This *Aviation Week & Space Technology* pilot recently flew the eighth production OH-58D to observe these capabilities. The 50-min. flight was conducted from Bell's production facilities here with Bell experimental test pilot Sam Boyer. It included low-level navigation to a nearby operating area and the tracking and simulated designation of another OH-58D serving as a target for our sensors.

Day/night helicopter

Key to the OH-58D's effectiveness as a day/night scout helicopter is the mast-mounted sight designed and developed by McDonnell Douglas Astronautics Co. in association with the Northrop Corp. Electro-Mechanical Div. The sight, mounted at the top of the rotor mast above the main rotor, comprises a telescopic TV camera for day observation, a forward-looking infrared system for night and day observation, and a laser rangefinder and target designator to mark targets for laser-guided weapons launched from other platforms including ground-based artillery, attack helicopters such as the Army/McDonnell Douglas AH-64 Apache, or fixed-wing attack aircraft.

The sight is gyro-stabilized and vibration-isolated, and provided surprisingly clear pictures of the surrounding environment on the cockpit displays associated with the system. Those displays, developed by Sperry Flight Systems, include two identical multifunction displays, one located in front of each crewmember, and a remote frequency display mounted at the upper center of the instrument panel.

The remote frequency display shows the frequencies selected on the five communications radios used in the OH-58D and provides control of communications security equipment.

The two multifunction displays provide an interface between the crew and the aircraft's visionic package. In addition to reproducing the images seen by the sensors in the mast-mounted sight, the multifunction displays also provide all necessary flight information, and mission status and subsystem monitoring.

Although the displays can be controlled using switches around the screens, the pilot can select the information he wants on his own display using switches on his cyclic control stick. This eliminates the need for the pilot to release the controls while he changes his display.

Boyer started the aircraft while I became familiar with the displays and their operation. Once the rotor was turning at its normal operating speed, Boyer aligned the Singer-Kearfott AN/ASN-137 Doppler navigation system that is part of the mission equipment package. The Doppler system and the Litton LR-80 attitude heading reference system (AHRS) installed in the aircraft provide a capability for reasonably precise low-level navigation, permitting the OH-58D to maintain a low profile throughout its mission.

With all systems operating, I was able to track several other helicopters in the area using a thumb switch on the copilot/observer's cyclic control stick to slew the mast-mounted sight. Slew rate of the sight was proportional to the pressure I applied to the thumb switch,

Army/Bell OH-58D Aeroscout will be used primarily for tracking and designating ground targets to be struck by artillery or helicopter and fixed-wing attack aircraft. McDonnell Douglas mast-mounted sight seen above the rotor system contains a daytime television sight, a day/night infrared imaging sensor and a laser designator/rangefinder.

Four OH-58Ds from the first production lot of 16 aircraft fly in formation during testing and training.

and it was possible to track the targets easily no matter what the rate at which they crossed our field of view. It was also possible to depress the thumb switch straight down to designate a target, after which the sight continued to track automatically.

Boyer conducted the first takeoff, and we lifted off to enter a trail position about 1.5 mi. behind the OH-58D that was serving as a target en route to a nearby training area. From that position, it was still easy to track the other aircraft with the mast-mounted sight in both the television and infrared modes.

The OH-58D is powered by a single General Motors Allison Gas Turbine Div. T703-AD-720 producing 650 shaft horsepower. This is a substantial increase over the 420-shp. Allison T63-A-720 engine used in the OH-58C, and there is a very comfortable margin of power available despite the fact that the expanded mission equipment package in the OH-58D adds approximately 890 lb. to the weight of the aircraft.

Normal takeoff weight of the OH-58D is about 4,150 lb., and maximum takeoff weight is 4,500 lb. Only about 65% torque was necessary to maintain a hover in ground effect on our initial takeoff with light winds and a temperature of 78F.

At 100 kt., the maximum allowable speed with the doors removed, the four-blade, soft-in-plane composite main rotor provided a smoother ride than earlier Bell two-blade rotor systems, and this made it all the easier to track the other aircraft or occasional ground targets.

We cruised at an altitude of 1,000 ft. with the AHRS inertial navigation system providing steering commands on Boyer's multifunction display to several waypoints he had entered into the system prior to

McDonnell Douglas mast-mounted sight above the main rotor of the OH-58D is gyro-stabilized and vibration-isolated. The 25-in.-dia. ball does not turn with the rotor, but rather is controlled directionally by the pilot or copilot/gunner. Television camera window is at left and infrared imaging sensor window is at right. Protrusion on support structure is a heat exchanger.

takeoff. The waypoints corresponded to identifiable objects on the ground, and it was easy to judge the accuracy of the system under these circumstances. There was very little deviation from these ground fixes as we made our way to the area that Bell uses for training and flight testing.

Once we entered the training area, Boyer landed the aircraft and I took the controls in the left seat to lift the helicopter into a hover. The OH-58D is equipped with a stability and control augmentation system (SCAS), and it has the solid feel of a much heavier helicopter in a hover.

In addition to the other modifications, the OH-58D incorporates an improved tail rotor capable of maintaining directional control in crosswinds up to 35 kt. This feature also allows quick, crisp turns in the hover.

After acclimating myself to the feel of the aircraft, I accelerated forward in nap-of-the-earth flight to begin looking for the other OH-

Instrument panel of the OH-58D includes large multifunction displays for the copilot/gunner, left, and for the pilot, right, as well as a remote frequency display at upper center of the panel showing frequencies selected on the aircraft's radios. Conventional electro-mechanical gauges and instruments are used only as a backup. Radar warning display is at far right.

Rear cabin of the OH-58D is filled with avionics boxes associated with the integrated target acquisition/designation system, precluding the carriage of passengers or cargo.

58D serving as our target. Using the switches on the instrument panel, I had selected infrared imagery from the mast-mounted sight to be shown on my multifunction display.

The VHF, UHF and high-frequency single-sideband radios in the OH-58D have the capability for secure communications by allowing messages to be sent to and printed out on the aircraft's multifunction display without the need for voice transmissions. While I was hovering the helicopter, Boyer had sent a secure message to the other OH-58D advising that we were ready to begin searching for it. The other helicopter, which was on a separate training mission, was to have sent a message in response advising that it was prepared to serve as our target once again. These messages must be entered into the system by the crew using a keyboard on the instrument panel.

Although the system normally works well, according to Boyer, in this case we were receiving no return message from the other aircraft. Through voice communications, the crew of the other aircraft confirmed its receipt of our message and advised they had responded.

We gave up after several additional tries, and Boyer later concluded that he may have been using an improper coding in attempting to receive the other aircraft's transmissions.

We continued our search using the mast-mounted sight while I maintained a slow cruise of 50–60 kt. at an altitude of 10–15 ft. This worked to keep us obscured from detection by the other aircraft, which had landed in a creek bed about 2 mi. away. The area in which we were operating included some low, gently rolling hills, and I kept the OH-58D between and below the tops of two of these as we conducted our hunt.

I brought the helicopter to a hover again as we neared a shallow valley where we knew the other OH-58D had been operating. Staying behind a low hill that separated us from the valley, I slewed the mast-mounted sight slowly from left to right until the image of the other aircraft appeared on my multifunction display. Although it was about 1.5 mi. away, the image was clear and easily identifiable.

The effectiveness of the mast-mounted sight in reducing the exposure of our own aircraft was underscored when I looked through the windscreen and could see nothing but blue sky over the top of the hill in front of us. The sight, which is only 25 in. in diameter, provides a distinct advantage over nose-mounted sights that require the entire helicopter to be exposed to locate a target.

Had this been an actual combat situation, I would have used the automatic target handoff system to digitally transmit target identification and location, including latitude, longitude and altitude, to ground and airborne attack elements. The target location is automatically computed through the use of the mast-mounted sight and control/display system subsystems, while target identification and suggested armament information can be manually entered through the use of the multifunction display keyboard.

As in many other recent aircraft programs involving integrated avionics systems, computer software development has been more of a challenge than anticipated. Familiarization and training of crews in the idiosyncrasies of these complex systems also have been more demanding than anticipated, and may have been the case here in our attempts to use the secure communication capabilities. Having completed the tracking exercise, I rolled the aircraft to the right and applied heavy right rudders in a "duck and run" maneuver that future scout and attack helicopter pilots are likely to use extensively to avoid detection or to evade return fire once they have completed an attack.

This sudden descent with an acceleration in sideward flight transitioning to forward flight is a relatively severe maneuver requiring large power and control margins. These are available in the OH-58D and allow such maneuvers to be completed without difficulty. Additional low-level flight on our return to the Bell production facility underscored again the agility of the OH-58D.

Bell began training Army instructor pilots, observers and maintenance personnel here in March in preparation for the fielding for the first operational OH-58D units. The service is preparing for a follow-on operational evaluation of the aircraft next year.

Results of the original operational testing of the OH-58D failed to show that the aircraft offered significant advantages over existing OH-58Cs in the scout and air cavalry missions identified for the aircraft. However, this was attributed more to the training of crews participating in the test rather than to any basic shortcomings in the aircraft. Outcome of the follow-on operational test could determine whether modification of all 578 aircraft will be funded or if the program will be stopped at fewer than 200 aircraft for the field artillery aerial observer role only.

Marine/Bell AH-1W demonstrates improved performance, flexibility

Robert R. Ropelewski/Ft. Worth
August 25, 1986

Substantial increase in power, a wider range of weapons and improved cockpit displays have been combined in the U.S. Marine Corps/Bell Helicopter Textron AH-1W SuperCobra to give Marine attack helicopter crews greater flexibility and lethality against increased threats on the ground and in the air.

The roles and missions of the AH-1 Cobra in both the Army and the Marine Corps have been expanding ever since it was originally conceived in the late 1960s as a close air support aircraft and as an armed escort for unarmed transport/assault helicopters. The AH-1W represents the latest evolution of this system with significant increases in performance and firepower.

The Marine Corps took delivery of the first production AH-1W here at Bell Helicopter's Ft. Worth plant late last March, and operational Ma-

First production Marine Corps/Bell Helicopter AH-1W SuperCobra takes off from Arlington, Tex., municipal airport, where Bell's flight test facility is located, for a demonstration flight. Aviation Week & Space Technology's *Washington bureau chief, Robert R. Ropelewski, is in the front cockpit, with Bell senior test pilot A. Monte Nelson in the rear cockpit.*

rine air units are scheduled to begin receiving their first aircraft this month (AW&ST Apr. 14, p. 109). This *Aviation Week & Space Technology* pilot recently flew a 1.1-hr. mission in the first production AH-1W to observe the upgraded capabilities incorporated in the aircraft.

Key to these upgrades is the installation of two General Electric T700-GE-401 turboshaft engines producing 1,690 shp. each at sea level—a 65% increase in installed power over the Pratt & Whitney Canada T400 Twin Pac powerplants used in earlier Marine Corps versions of the Cobra. This additional power allows the aircraft to carry a wider variety and greater number of weapons while still meeting Marine Corps requirements for a hover-out-of-ground-effect capability with 14,000 lb. gross weight up to 3,000 ft. at 91.5F, or at sea level in temperatures up to 103F.

Distinctive features

Other features distinguishing the AH-1W from earlier Cobra versions include:

- Provisions to carry and launch Rockwell International Hellfire laser-guided antitank missiles and Ford Aerospace/Raytheon AIM-9L Sidewinder short-range air-to-air missiles. The SuperCobra also can carry and launch the Hughes Aircraft Co. TOW optically tracked, wireguided antitank missile, but the AH-1T Cobra operated by the Marines has this capability as well.
- Upgraded transmission rated at 2,032 hp., compared with 1,970 hp. for the earlier AH-1T and 1,250 hp. for the AH-1J.
- Crashworthy fuel system including honeycomb fuel cell panels with fire suppressant powder inside.
- Kaiser head-up display for the pilot in the rear cockpit with flight guidance and weapon delivery symbology.
- Cockpit lighting system compatible with night vision goggles.
- New avionics package including dual AN/ARC-182 UHF/VHF communications radios operating in three frequency spectrums, and a countermeasures package comprising an AN/APR-39(V)1 radar detector, AN/APR-44(V)l radar warning system, AN/ALQ-144 infrared jammer and AN/ALE-39 chaff/flare dispenser.
- A vibration suppression system that uses a hydraulic damper beneath the cockpit floor in the nose of the aircraft to counter main-rotor-induced vibrations in the fuselage. Ride quality and weapon-aiming precision were noticeably improved with this system operating.

The *Aviation Week* evaluation flight in the SuperCobra was flown from Bell Helicopter's Arlington, Tex., flight test facility. Bell senior test pilot A. Monte Nelson provided an initial demonstration of the aircraft's performance and flying quality from the rear cockpit—the pilot's station—of the AH-1W until I became familiar enough with the controls and displays in the front cockpit to operate the aircraft from that position.

Marine Corps and Bell program managers on the AH-1W said they considered integrating the aircraft's avionics and weapon delivery systems through a multiplex bus, but decided against this because of cost and the unavailability of compatible systems meeting the Marine Corps' mission requirements.

Component location

One result of this decision is that fire control and ECM system components are widely dispersed around the pilot's station in the rear cockpit. Controls for the AIM-9L Sidewinder air-to-air missiles are mounted on the sloping left glare shield, for example, with gunnery accuracy control just above the Sidewinder control panel to the left and slightly below the head-up display.

The radar warning indicator is mounted to the right of the HUD, and controls for the chaff/flare dispenser are positioned just below the radar warning indicator on the sloping right side of the glare shield. Other weapon controls for guns and rockets are located at the lower center of the front instrument panel.

In combat, the pilot is expected to have primary responsibility for the aircraft's guns, rockets and AIM-9 missiles. The weapons operator in the front cockpit is primarily responsible for the TOW and Hellfire missiles carried by the helicopter. The AIM-9 system can take immediate priority over all other weapons on the AH-1W in order to respond quickly to air-to-air threats.

The front cockpit accommodating the gunner's station is somewhat better organized with a Canadian Marconi-developed panel on the left side of the front instrument panel consolidating the control and display functions of both the TOW and Hellfire missiles. Primary flight instruments are located across the top of the front panel, and engine instruments are grouped to the far right of the panel. A head-down TOW missile telescopic sight extends upward from the lower center of the front panel. A mechanical side-stick cyclic control handle is mounted on the right console in the front cockpit, permitting a copilot/gunner to operate the aircraft from his station when necessary.

The well-equipped cockpit of the AH-1W includes AIM-9L Sidewinder air-to-air missile control panel, head-up display and radar warning indicator above the glare shield, as well as gun and rocket control panels at lower center of front instrument panel (behind stick). Built-in work stands and other accessibility features have been incorporated in the AH-1W SuperCobra to improve maintainability and reliability.

Nelson started the T700 engines one at a time, leaving the rotor brake engaged until both engines were running at ground idle. The engines can be started with the rotor brake on or off. Normally, the brake would be used only to avoid excessive rotor blade flapping at low rpm in high winds. In this case, the prevailing wind was reported to be 16 kt. and was not a significant factor.

Maximum allowable measured gas temperature (MGT) is 960C during engine start, but neither engine exceeded 650C during the starts. Each engine reached idle about 20 sec. after the starter was engaged. Nelson released the rotor brake and the two-blade, semi-rigid rotor accelerated to flight rpm.

Primary system checks

Approximately 1 min. was required to accomplish primary system checks, and we then engaged the vibration suppression system (VSS).

The system is located directly beneath the floor of the copilot/gunner's station, and there was a chattering vibration beneath my feet for a second as the hydraulic damper associated with the system was actuated.

Aircraft weapons configuration for this flight included two Hellfire missiles on the right weapons pylon, four TOW missiles on the left pylon, chaff dispensers on top of each pylon and a three-barrel 20-mm. cannon in the chin turret. This, plus 1,500 lb. of fuel, brought the aircraft to a typical mission weight of about 13,125 lb. as we prepared for takeoff.

Addition of new weapon-carrying capabilities with each iteration of the Cobra series of attack helicopters has resulted in a steady growth in both the empty weight and maximum gross weight of the aircraft. Empty weight of the AH-1W—10,140 lb.—is higher than the 10,000-lb. maximum gross weight of the AH-1J, the first Cobra to be flown by the Marine Corps.

The AH-1J had an empty weight of 6,610 lb. This increased to 8,553 lb. when a TOW missile capability was added to the Cobra configuration to produce the AH-1T. Maximum gross weight of the AH-1T is 14,000 lb. Addition of the Hellfire missile capability in the AH-1W

Hughes Aircraft Co. TOW missile launch tubes are hung from weapon pylon at lower left, and a chaff dispenser is mounted on top of the pylon.

raised the empty weight of this version to 10,140 lb. and its maximum gross weight to 14,750 lb.

With an outside air temperature of 82F/28C and a field elevation of 630 ft., the AH-1W was well within its operating limits as Nelson lifted the aircraft into a hover. About 65% torque on the main transmission was required to hold the helicopter in a 10-ft. hover.

The SuperCobra's environmental control unit kept the temperature in the cockpit comfortably cool despite the hot midday sun shining through the canopy. Turning off the ECU would have provided a few percent more hover power if that had been necessary.

Nelson tipped the nose forward to accelerate and increased power to 100% torque to establish the aircraft in a 70-kt. climb at 2,000 fpm. Measured gas temperature on both engines was about 750C under these circumstances. At an altitude of 1,000 ft., he rolled the throttle to idle on the No. 1 engine to simulate an engine failure.

Engine torque is limited to 79% for 2.5 min. with one engine inoperative, and to 74% for maximum continuous operation. With 74% torque on the No. 2 engine and a sustained airspeed of 70 kt., the aircraft stabilized at a climb rate of about 1,000 fpm. The Marine Corps defined a requirement for a single-engine rate of climb of 800 fpm at 3,000-ft. altitude on a hot day, a requirement that has been exceeded with the AH-1W.

Nelson transitioned to level flight at 1,600 ft., still maintaining 74% torque on the No. 2 engine. The aircraft stabilized at a cruise speed of 110 kt. The environmental control unit was still operating at this point. Turning it off would have provided a few percent additional power for an increase of several knots in airspeed. In the AH-1W, as in earlier versions of the Cobra, the unit shuts off automatically in the case of an actual engine failure.

Outside air temperature was still 28C at 1,500 ft. when Nelson brought the No. 1 engine back to life again to resume normal twin-engine operations. With the power set at 100% torque on the main transmission and 750C measured gas temperature on each engine, airspeed increased to 150 kt.—about what the AH-1W performance charts indicate for those conditions. The Marine Corps requirement was for a maximum speed of at least 152 kt. at sea level on a standard day with a maximum gross takeoff weight.

Torque limiter

An automatic torque limiter is incorporated in the aircraft to prevent total torque from exceeding 105–106% during twin-engine operations. For single-engine operations, there is no total torque limit, but a maximum measured gas temperature of 917C is permitted for no more than 2.5 min. on the operative engine.

Vibration levels were still relatively low at 150 kt., although they had been intensifying slowly as our speed increased above 120 kt. Nelson turned off the vibration suppression system momentarily, and there was a significant increase in the amplitude of the vibrations. He turned the VSS on again and the vibrations subsided.

At 150 kt. with a light load and an aft center of gravity, the AH-1W rotor system has a constant two-beat-per-revolution vibration of less than 0.1g with the vibration suppression system engaged. With the VSS off, the vibration intensity more than doubles but is still less than that of AH-1T, which is not equipped with a VSS.

Head-down weapon aiming sights such as the TOW sight used on the Cobra are difficult to use with precision on helicopters at high

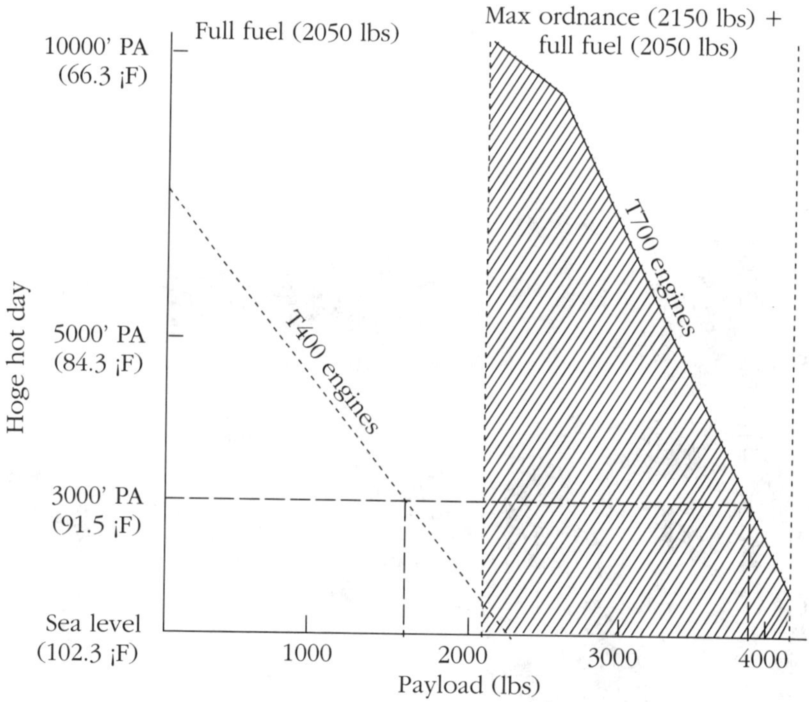

Weapons payload is nearly doubled and high/hot operating capabilities have been substantially increased in the T700-powered AH-1W compared with the T400-powered AH-1T, as this Naval Air Systems Command performance comparison indicates. The chart shows hot-day hover-out-of-ground-effect performance for the two aircraft.

speeds because the vibration levels cause the operator's head and eyes to vibrate against the sight itself. The VSS on the AH-1W appears to reduce this problem to a great extent, and should improve the precision guidance of the aircraft's TOW missiles.

We climbed to 3,000 ft., where Nelson reduced power to about 65% torque and rolled the aircraft into a 25-deg. dive for a simulated weapon delivery run. We accelerated to 170–175 kt. during the dive, but the vibration suppression system kept vibrations at a relatively low level until Nelson applied approximately 2.5g to pull out of the dive. At that point the vibration levels reached an intensity that exceeded the capabilities of the VSS, and there was a noticeable increase in two-per-rev shaking in the cockpit for a few moments.

Visibility is excellent from the SuperCobra cockpit, but we did not discern a Cessna 182 that was approaching from about our 4 o'clock position at the same altitude. It was not until we were rolling left to enter the dive that both Nelson and I noticed the Cessna a few hundred feet away as it banked sharply to the right to avoid us.

Normal armament configuration for the AH-1W SuperCobra includes a three-barrel 20-mm. cannon mounted in a chin turret. The cannon can be controlled by either the pilot or gunner. TOW missile telescopic sight can be seen in the front cockpit of the helicopter.

Gurney flap, a fixed rectangular plate at the rear of the vertical stabilizer on the AH-1W, was adopted to counter the effects of rotor blade flapping at high altitudes.

I took the controls after the simulated weapon delivery run to repeat most of the same maneuvers demonstrated by Nelson, with similar performance results. At 1,700 ft. and 120 kt., Nelson again reduced the No. 1 throttle to idle to simulate an engine failure. Still on the controls, I adjusted the collective pitch control lever slightly to establish the maximum continuous torque setting of 74% on the remaining engine. This resulted in a measured gas temperature of 800C on the operative engine and a slow but steady acceleration to 135 kt.

High-altitude performance of the AH-1W was evident when I took the aircraft to 8,000 ft. and entered a hover. Aircraft weight at that point was approximately 12,700 lb. and the outside air temperature was 17C (63F). The SuperCobra maintained altitude in the hover with the power at 85% torque. Performance charts for the AH-1W indicate the aircraft could have hovered out of ground effect at 10,000 ft. or higher at that weight. Alternatively, we could have hovered out of ground effect at 8,000 ft. with an aircraft weight of up to 13,500 lb. At much lower altitudes, the excess power available in the SuperCobra leaves an impression of substantial agility. During nap-of-the-earth flying and bob-up maneuvers from a hover behind defilade to point and shoot at simulated targets, the aircraft demonstrated excellent stability and controllability characteristics. This was true whether we were maneuvering around obstacles while navigating at low altitude or popping up behind a line of trees from a 5-ft. to a 50-ft. hover to simulate firing at a target. Large margin of safety in tail rotor controllability allowed us to turn and dart quickly to a new position after these simulated firings.

The aircraft also handled easily in the landing pattern at the Arlington flight test facility, where I performed a series of approaches to a hover before making a final landing. The excess power available provided positive control in arresting steep approaches or in preventing any settling after a nose-high flare at the end of a fast approach.

One minor distraction encountered in these approaches was the lower mechanical advantage of the side-stick cyclic control in the front cockpit compared with the pilot's conventional center-mounted cyclic control stick in the rear. In the final few knots of each approach to a hover, the effect of the tail rotor on the vertical fin became more pronounced and tended to push the tail of the aircraft to the left. This required more left rudder and a firm push to the right on the stiff cyclic control at the end of each approach. Considering that the aircraft would not be routinely operated from the front cockpit, however, this was a minor annoyance.

Electronic engine fuel controls on the T700 engines were exceptionally responsive to collective lever movements throughout the flight, and even the most abrupt changes in collective pitch generated less than a 1% change in rotor rpm. Normal operating range for the rotor is very narrow—99–100%. Voice, tone and warning light advisories are activated whenever rotor rpm drops to 97% or increases to

103% or higher. The only time I saw or heard these warning signals was on startup and shutdown.

Following our final landing and a 2-min. cool-down period on the T700s, Nelson brought the engines to idle and applied the SuperCobra's rotor brakes when rotor rpm slowed to 47%. The brakes took hold at 45% rpm and quickly stopped the rotor. The rotor brakes can be applied at 60% rotor rpm or less.

The Marine Corps currently has six helicopter light attack squadrons each operating 12 AH-1J and AH-1T Cobras. The existing order for 44 new AH-1Ws delivered at a rate of two aircraft a month over the next two years is "not an attempt to expand the attack force," according to Col. Marvin F. Pixton, 3rd, program manager for Bell UH-1 and AH-1 and Rockwell OV-10 programs within the Naval Air Systems Command.

Instead, he said, the initial 44-aircraft AH-1W buy is an effort to compensate for attrition of the older AH-1S. In addition to the 44 new aircraft, the Marine Corps is planning to modify 44 of its AH-1Ts to the AH-1W configuration and is also seeking Fiscal 1987 funding to order 34 more new AH-1Ws (AW&ST Apr. 14, p. 109). Conversion of each of the existing AH-1Ts is expected to cost about 70% of the price of a new AH-1W, Pixton said.

In the meantime, program officials have given a high priority to the incorporation of a night targeting capability in the AH-1W, combining a forward-looking infrared system, individual crewmember night vision goggles and a laser designator/ranger. A request for proposals is expected to be issued before the end of the year. Another high-priority requirement for the SuperCobras is the addition of a third wing station on each side for air-to-air weapons such as the AIM-9 Sidewinder or the General Dynamics Stinger lightweight missile.

First production AH-1W navigates at low altitude armed with four Rockwell International Hellfire laser-guided antitank missiles under its left weapons pylon, and a three-barrel 20-mm. cannon in the nose.

CH-47D offers upgraded performance

Donald E. Fink/Philadelphia
July 18, 1985

Army/Boeing Vertol CH-47D Chinook, the latest version of the company's twin-engine, tandem rotor medium-lift helicopter, combines improved handling qualities with increased performance that will make it a key element in the U.S. Army's operational battlefield inventory through the year 2000.

Army has a 10-year Chinook upgrade program that includes orders for 436 CH-47Ds, which are modernized remanufactured versions of earlier CH-47A, B and C helicopters. Deliveries of this version are expected to run through 1991. The Army also is planning an additional buy of 142 newly manufactured CH-47Ds, with deliveries that will extend into mid-1995.

CH-47D Chinook taxis at a speed of about 5 kt. in the Elk River during water operations conducted on the second part of a two-flight evaluation made by Aviation Week & Space Technology *pilot Donald E. Fink. Reducing power to ground idle cuts forward speed in the water to about 2 kt.*

Helicopter tilts forward to accelerate following liftoff from the river.

Performance of the CH-47D was evaluated here recently by this *Aviation Week & Space Technology* pilot during two flights, totaling 3 hr. 30 min., in the No. 9 production machine. It is the 12th remanufactured helicopter delivered from Boeing's CH-47D production line. The first three were prototypes.

The *Aviation Week* evaluation flights were made with a three-man Boeing Vertol crew that included J.R. Mecklin, CH-47 project test pilot; Larry J. Hartman, flight test engineer, and Frank W. Cross, flight test mechanic. Mecklin occupied the right seat during the flights, Hartman functioned as flight engineer and cargo master and Cross handled the cargo sling hookups during external load tests.

The primary impression I gained from the two flights, which included a series of orientation flying maneuvers, five external cargo lifts and five water landings, was that the CH-47D is easy to fly, even for a pilot whose helicopter experience is limited mainly to smaller single-rotor machines.

The enhanced handling qualities of the CH-47D are due principally to a series of improvements made in the flight control and stabilization systems, according to Mecklin. The advanced flight control system in the CH-47D uses four-axis stabilization and trim hold to give the pilot airspeed and pitch attitude hold, bank angle hold, heading hold and altitude hold. The altitude hold function is governed by a radar altimeter for hovering and low-altitude very-low-speed flight, and by a barometric altimeter for high-level cruise.

"The barometric altimeter will hold altitude within ±25 ft. during cruise in smooth air," Mecklin said. "The radar altimeter holds ±2–4 ft. at hover altitudes up to about 50 ft. It functions up to 1,500 ft., but it's designed for low-level work. I would not expect it to be very accurate at higher altitudes."

The advanced flight control system has essentially the same capabilities as an autopilot. However, it is a duplex system instead of a self-monitoring triplex system and has limited authority in all three axes, so it cannot be used for long-term hands-off flying. "But it is possible to let the system do much of the work in flying the helicopter," Mecklin said.

The system provides four control modes a pilot can use to change heading, he said. The cyclic stick can be moved out of the detent to achieve the desired bank angle and then released. The bank angle hold function will maintain the selected turn rate until the stick is moved to reestablish level flight.

A coolie hat Vernier switch on the cyclic stick head also can be used to "beep" the control system into 30–35-deg. bank turns. Beep trim is available for pitch attitude and airspeed control as well.

Turns also can be accomplished by selecting a desired heading on the radio magnetic indicator and letting the advanced flight control system turn the helicopter into the new heading. In this mode, a standard rate turn is commanded at airspeeds up to 120 kt., Mecklin said. At higher speeds, turns are limited to 20-deg. bank angles. The helicopter also can be flown manually by moving the cyclic stick and overriding the electronic system.

Improvements in the performance of the CH-47D result from the use of Avco Lycoming T55-L-712 turboshaft engines, replacing earlier versions of the T55 used in the CH-47A, B and C models. The CH-47D main transmission also has been uprated to 7,500 shp. and advanced composite material has been used for the new rotor blades. The three-bladed rotor design from earlier Chinooks is retained on the CH-47D. The performance gain attributed to the composite blades is 4–5%, Mecklin said. Their primary advantages are their unlimited lifetimes and the ease with which they can be repaired in the field.

The T55-L-712 engines have a normal continuous rating of 3,000 shp. as installed in the CH-47D. They also can be operated for 30 min. at an intermediate rating of 3,400 shp., for 10 min. at a maximum rating of 3,750 shp. and for 30 min.—cumulative time—at an emergency rating of 4,500 shp. A timer is activated whenever the turbine inlet

(continued on page 62.)

CH-47 Characteristics Compared

	CH-47A	CH-47B	CH-47C	CH-47D
Internal Cabin Dimensions				
Length (in.)	366	366	366	366
Width (in.)	90	90	90	90
Height (in.)	78	78	78	78
Usable Fuel Capacity (gal.)	564	564	1,042	1,042
Transmission Rating (shp.)	4,970	4,970	6,000	7,500
Avco Lycoming Engines	T55-L-7C	T55-L-7C	T55-L-11	T55-L-712
Ratings				
(shp., uninstalled, sea level)				
Max. Continuous	2,400	2,400	3,000	3,000
Intermediate (30 min.)	2,650	2,650	3,400	3,400
Takeoff (10 min.)	2,850	2,850	3,750	3,750
Emergency	—	—	—	4,500
External Cargo Load				
Capacity (lb.)				
Forward	—	—	—	17,000
Aft	—	—	—	17,000
Center	16,000	20,000	20,000	26,000
Forward, Aft Dual Point Load	—	—	—	25,000

Boeing Vertol CH-47 Helicopter Models

	CH-47A	CH-47B	CH-47C	CH-47D
Capacity				
Empty Weight (lb.)	18,288	19,676	21,586	23,093
Fixed Useful Load (lb.)	1,382	1,382	1,400	1,403
Design Gross Weight (lb.)	28,550	33,000	33,000	33,000
Maximum Gross Weight (lb.)	33,000	40,000	46,000	50,000
Rotor Speed (rmp.)	230	225/230	235/245	225
Usable Fuel Capacity (lb.)	3,666	3,666	6,773	6,695
Performance (33,000 lb./std. atmos.)				
Maximum Cruise Speed at Sea Level (kt.)	110	155	161	158
Maximum Rate of Climb at Sea Level in 30 Min. (fpm.)	2,040	2,080	2,590	3,100
Hover Out of Ground Effect Ceiling in 10 Min. (ft.)	8,800	10,100	13,600	17,200
Service Ceiling (30 min. rating) With One Engine Inoperative (ft.)	2,000	—	8,500	12,800
Mission Capability				
External Payload Mission (sea level/59F)				
Takeoff Gross Weight (lb.)	33,000	39,450	44,400	50,000
Total Mission Fuel (lb.)	2,267	2,545	2,848	2,818
Outbound Cruise Speed (kt.)	100	100	100	126
Inbound Cruise Speed (kt.)	130	132	137	135
Payload (lb.)	11,063	15,847	18,566	22,686

External Payload Mission (2,000 ft./70F)				
Takeoff Gross Weight (lb.)	33,000	38,200	42,950	50,000
Total Mission Fuel (lb.)	2,108	2,464	2,698	2,738
Outbound Cruise Speed (kt.)	100	100	100	112
Inbound Cruise Speed (kt.)	120	140	140	137
Payload (lb.)	11,222	14,698	17,266	22,766
External Payload Mission (4,000 ft./95F)				
Takeoff Gross Weight (lb.)	31,100	32,500	40,700	42,900
Total Mission Fuel (lb.)	1,988	2,222	2,542	2,548
Outbound Cruise Speed (kt.)	92	100	100	101
Inbound Cruise Speed (kt.)	103	144	146	140
Payload (lb.)	9,442	9,220	15,172	15,856

temperature on an engine reaches or exceeds 900C—the emergency power operating range—and stops when it falls below that level. When the timer on an engine is activated, that engine must be inspected before the timer is reset.

After a walkaround inspection and a cockpit briefing at Boeing Vertol's Ridley Park flight facility in suburban Philadelphia, Mecklin demonstrated the CH-47D's self-contained start procedure. The helicopter has a Solar Turbomach auxiliary power unit that is energized by a hydraulic starter motor. The APU automatic start procedure is activated by a three-position switch that has off, start and run positions. Once running, the APU is used to drive the engine starter and to provide power for systems checkouts on the ground.

The main engine start procedure is relatively simple and consists of moving the starter switch to the motor position until the engine reaches 10–15% N_1 speed. At this point, the engine power lever is moved from the stop to the ground position, which starts fuel flowing to the engine. The starter' switch then is moved to the start position and engine light is monitored on the turbine inlet temperature indicator. During a normal start, the temperature peaks at about 550C.

The starter is left engaged until the engine reaches 50% N_1 speed, at which point it is switched off. The engine usually stabilizes at between 60–63% N_1. When both engines are started, the power levers are moved to the flight position and the engines are beeped from flight idle to 100% N_2/N_R. The engines are left at 100% for all normal operations.

Automatic clutches engage the two counter-rotating rotors as soon as torque is exerted by the engines. There are no rotor brakes on the U.S. Army's version of the Chinook. Standard rotor operating speed is 225 rpm.

When the engine start and systems checks were completed, the tower at Boeing's flight facility cleared us for taxi. The CH-47D has fixed landing gear units at each corner of the fuselage—two double-wheel bogies in the front and two single-wheel bogies in the rear.

The rear units swivel to facilitate turning during ground taxi, and the right unit has an electric power steering system for tight maneuvering with heavy loads. The rear wheels are locked for takeoff and landing.

Ramp weight at engine start was 31,000 lb., including 7,000 lb. of fuel. Despite its size, I found the CH-47D easy to taxi, whether using cyclic and rudder pedal inputs or the rear wheel power steering. On soft terrain, the Chinook also can be taxied on the rear wheels by locking the aft gear units and lifting the nose, Mecklin said.

After I did a series of liftoffs and touchdowns and hovering turns to get an initial feel for the CH-47D's handling characteristics, especially its lack of rotor torque forces, Mecklin took the controls for the first departure from the Boeing ramp. William P. Jones, director of helicopter programs at Boeing Vertol, rode in the observer's jump seat for the first portion of the flight.

Standard departure profile is southeast across the Delaware River to New Jersey via a corridor with a ceiling of 300 ft. This keeps helicopter traffic below the floor of the terminal control area (TCA) for Philadelphia International Airport, which is located a few miles north of the Boeing Vertol plant. Once across the river, the TCA floor rises to 600 ft., 1,500 ft., 3,000 ft. and finally 4,000 ft.

Shortly after Mecklin made a turn to fly southwest along the New Jersey side of the Delaware, the master caution light illuminated, calling attention to a chip detector light on the No. 2 engine. Power on that engine was reduced and we returned to the Boeing Vertol ramp. The chip detector was pulled and found to have accumulated enough metallic "fuzz" to illuminate the chip warning light, but no chips were present.

Following the second departure from Ridley Park, I flew the CH-47D southwest to the Woodstown, N.J., VOR for a Doppler navigation system check and continued southeast for a series of maneuvers at an altitude of 3,000 ft. to evaluate the helicopter's handling characteristics. At a gross weight of slightly over 30,000 lb. and an indicated airspeed of 140 kt., the Chinook handled comfortably in maneuvering turns with bank angles up to 60 deg. The red line airspeed is 170 kt.

At its maximum gross weight of 50,000 lb. and with an internal load, the CH-47D is limited to 45-deg. bank angles. Bank angle limits with external loads vary, depending on the type of load and sling rigging used.

Performance of the advanced flight control system also was evaluated in maneuvers that demonstrated the effectiveness of the bank angle, pitch angle, differential airspeed and the heading hold features. Collective control inputs and power setting changes were offset by the heading hold function with no tendency toward hunting in the yaw axis.

To demonstrate the system's effectiveness in handling an engine failure, Mecklin pulled the power lever on the No. 2 engine to ground idle and I detected no yaw tendency that had to be corrected. Rotor rpm began to degrade, but that was offset by using the beep trim to increase power on the No. 1 engine.

"A lot depends on the gross weight at which you're operating when you pull an engine or lose one," Mecklin said, "but I haven't seen the rotor rpm go much below the middle of the lower yellow band—about 90%. It's possible to recover from that by beeping up the good engine."

Mecklin restored power to the No. 2 engine and switched off one channel of the advanced flight control system. There were no flight attitude excursions, but Mecklin said the flight manual calls for reducing airspeed to 120 kt.

Mecklin then shut off the second advanced flight control system channel, and the CH-47D remained flyable without an undue increase in pilot workload, but it did require what he termed "increased care and attention."

"With the AFCS shut off, it's possible to accelerate back up to 140 kt., but it's a good idea to stay at 120 kt," he said.

Army testing

Autorotations were not simulated during the flight evaluation, but were included in the Army flight test program. The procedure involved dropping the collective control and beeping both engines off-line, Mecklin said. Two different airspeeds were used—70 kt. for best rate of descent and 100 kt. for best angle of glide—and rates of descent varied from 2,100 fpm at a gross weight of 30,000 lb. to 3,000 fpm at 50,000 lb. Full landing autorotations were made at touchdown speeds of 35–40 kt.

After completing the airwork demonstrations, we returned to the Boeing Vertol ramp to drop off Jones. My approach to the ramp was

Army/Boeing Vertol CH-47D helicopter touches down in the Elk River during one of five water landings made as part of the flight evaluation.

too fast and resulted in a slight overshoot of the preferred landing spot. Mecklin helped correct the approach and we came to a hover and landed.

Our third departure from the Ridley plant was for the Greater Wilmington Airport in Delaware, where Boeing Vertol has its flight test center. By this time, I had developed a better feel for the proper approach technique, and my approach to the intersection of the main runways was much smoother.

Five cargo sling loads of 4,000-lb. Kirksite blocks—a high-density mixture of lead and babbitt—and a ballasted 8.5-×-8-×-20-ft. container van had been assembled at the edge of the parking ramp for the external loads demonstrations.

The first load included separate slings on all three of the CH-47D's centerline cargo hooks, with 8,000 lb. attached to the forward and aft hooks and a single 4,000-lb. block on the center hook. Each of the quadruple strap slings was 14 ft. long.

Mecklin demonstrated the pickup technique for the first load, which involved setting the radar altimeter for a 15-ft. hover and maneuvering the helicopter over the prescribed load. Hartman provided final positioning commands via the intercom, while Cross handled the sling hookups beneath the helicopter.

Liftoff with the 20,000-lb. external load was smooth and did not exceed the CH-47D's 50,000-lb. maximum gross weight limit because we had burned sufficient fuel. Prototype Chinooks have been test flown at a maximum gross weight of 56,000 lb., Mecklin said, and the company is offering the Army the option of increasing operational gross weight to 54,000 lb.

After lifting the 20,000-lb. load, Mecklin flew southeast across the Delaware River to a marshland area, where I flew a series of maneuvers to evaluate the helicopter's handling qualities with a large free-swinging external load. Longitudinal and yaw forces could be felt feeding back through the Chinook airframe at 120 kt. as the loads shifted and swung on the cargo hooks, but these were canceled by the advanced flight control system and did not have to be countered with control inputs. The feedback forces became more pronounced when I flew the helicopter through 30-deg. bank turns but not to the point of adding to the pilot workload.

We returned the 20,000-lb. triple-sling load to Greater Wilmington Airport and lifted the 4,000-lb. block on the aft hook to evaluate the CH-47D's performance with an aft c.g. load. I found the radar altimeter to be an excellent aid in maintaining a constant hover height, enabling me to concentrate on achieving and holding

Water cascades from sides of the CH-47D as it lifts vertically from the Elk River at the conclusion of water operations conducted during the flight evaluation. Landing and take-off techniques for water operations essentially are the same as those used on land.

the proper position over the load. This lighter load was handled easily in flight at speeds up to 140 kt. and 30-deg. bank turns.

Block combination

The third load lifted consisted of the 8,000-lb. block combination on the forward cargo hook. In this configuration, the longitudinal and yaw forces began to increase substantially at 130 kt. Mecklin advised slowing to a cruise of 120 kt., at which point the feedback disappeared.

For the fourth load, 20,000 lb. of Kirksite blocks were attached to the center hook. While we were carrying this load at a cruise speed of 110 kt., Mecklin pulled the No. 1 engine power lever to ground idle to simulate engine loss. We cut airspeed to 90 kt., which Mecklin maintained by beeping the No. 2 engine to 104% torque, still below emergency power.

The fifth load lifted was the container van, which was loaded with Kirksite blocks and 10-ft. rail/roller sections, bringing its total weight to 18,400 lb. This load was rigged with a 16-ft. sling on the forward hook, a redundant 20-ft. safety sling on the center hook and a 14-ft. sling on the aft hook.

With this rigging, which kept the van fairly streamlined with the direction of flight, we were able to cruise at 120–125 kt. When this type of load is attached only to the center hook, it can be carried at a maximum of 40–50 kt., Mecklin said.

After returning the final load to Wilmington, we landed on the ramp for a rest stop and refueling. There were 990 lb. of fuel remaining in the Chinook's fuel tanks, which are housed in large sponsons on each side of the fuselage. Slightly over 6,000 lb. of fuel were added using the Chinook's single-point refueling system located in the right sponson.

Ramp weight at engine start for the second flight again was 31,000 lb., and Mecklin demonstrated a high-rate power climb from a hover. Rate-of-climb was 2,000 fpm at 70% indicated torque.

After departure from Wilmington, we flew southwest to Port Herman, Md., for the water landing demonstrations in the Elk River. A 12-kt. wind was blowing from 270 deg. at the water landing site, which is where the Bohemia River flows into the Elk River. The landings were made on a heading of about 230 deg., with a slight crosswind from the right.

Mecklin made a couple of low-level survey runs over the landing site and then demonstrated the water landing technique, making the first touchdown from a hover. The approach to hover and touchdown are identical to those used on land. Heavy spray was kicked up by the rotors just prior to touchdown, but it was easily cleared from the windshield with the Chinook's two-speed wipers.

The Chinook settled into the water with a forward speed of about 5 kt. because of the forward tilt of the rotor shafts. Pulling the power levers to ground idle reduced this to about 2 kt. Bringing the rotors back up to speed, we taxied on the water, using cyclic inputs coupled with rudder pedals to do a series of turns. Slight back pressure had to be held on the cyclic to raise the nose and reduce stress on the lower forward windows. The river surface had a slight wind-induced chop and currents where the two river flows merged, but these caused no problems in handling the helicopter on the water's surface.

The procedure for taking off from the water also is the same as that used on land. The windshield wipers were needed only briefly to clear the spray. Four more water landings were made, including several running landings at a forward speed of about 20 kt. The procedure used for the latter is essentially the same as that used for a single-engine touchdown on land. A total of 20 min. was spent in the water.

After the last water takeoff, I flew back to the Woodstown VOR for a final Doppler navigation check. The system had not been updated since leaving Wilmington 45 min. earlier, but was only 2 km. off over the Woodstown VOR. From Woodstown, I flew northeast to the Delaware crossing corridor and made the final approach back to Ridley Park.

Apache to provide night/bad weather capability

Robert R. Ropelewski/Yuma, Ariz.
September 13, 1982

Initial deliveries of the Hughes Helicopters AH-64 Apache advanced attack helicopters to the U.S. Army in early 1984 will mark a major step toward a true night/adverse weather antitank capability for that service—a capability the Army believes will be essential in future high-threat European and Middle East environments.

Effectiveness of the integrated day/night/adverse weather navigation/attack system of the AH-64, along with the improved performance of the aircraft with its uprated 1,690-shp. General Electric T700-GE-701 turboshaft engines, has been acknowledged by Army pilots who have evaluated the Apache over the past year. These same characteristics were observed by this *Aviation Week & Space Technology* pilot in a recent series of demonstration flights in the aircraft at the Army's Yuma Proving Ground, where several years of AH-64 flight testing are coming to an end.

Army/Hughes Helicopters AH-64 attack helicopter prototype No. 6 flares for landing near completion of Aviation Week & Space Technology *evaluation flight in the aircraft. This was one of two flights in the AH-64 by Los Angeles bureau chief Robert Ropelewski, and concentrated on sensor and weapons systems aspects of the aircraft.*

Although there is still a lot of skepticism in some sections of industry and government over the survivability of the Apache in a high-threat environment and its reliability and maintainability in the battlefield, Army pilots and maintenance personnel who have assessed the aircraft in operational testing said it exceeds the Army's specifications in almost all respects and is an improvement over previous generations of attack helicopters in the Army inventory.

Three flights

My assessment of the Apache consisted of three separate flights using the fifth and sixth AH-64 prototypes, with one of those flights concentrating on weapon system operation in the front cockpit of the aircraft and the other two flights focusing on aircraft operation and performance from the front and rear cockpits.

Operation of the AH-64 as a complete weapons system is intended to be a two-man task, with the pilot, who sits in the rear cockpit, attending to navigating and flying the aircraft while the copilot/gunner concentrates on delivering an assortment of weapons on target from the front cockpit. Doppler navigation system, day/night vision systems, laser ranging/designating/tracking and digital fire control computer have been integrated in such a manner that they minimize the need for communications between the two cockpits while still allowing the crew to function as an effective team to accomplish the mission at a relatively low workload.

Weapons system aside, the AH-64 is a smooth and powerful machine with relatively wide performance margins for altitude and temperature extremes. With a typical mission gross weight of around 16,000 lb., the aircraft's two 1,690-shp. T700-GE-701 engines provide enough power to get the aircraft into the air without the need to make rolling takeoffs on a runway until a minimum flying speed is attained—a practice many Vietnam-era gunship pilots will recall.

The Army adopted the higher-powered T700-GE-701 engines a year ago to replace the 1,540-shp. T700-GE-700 engines originally used in the AH-64 prototypes. Purpose of this was to meet a newly evolved Army high/hot performance requirement for a minimum vertical rate of climb of 450 fpm at 4,000 ft., 90F, at intermediate rated power with eight Rockwell International Hellfire antitank missiles and 320 rounds of 30-mm. ammunition, and fuel for a 1-hr. 50-min. mission. Recently completed flight tests at the Army Flight Test Activity at Edwards AFB, Calif., demonstrated that the aircraft can climb at 860 fpm under those circumstances. This will allow carrying additional payload and/or fuel while still meeting the minimum climb requirement.

These high/hot performance characteristics were evident in my three flights in the AH-64, which were conducted at altitudes from sea level to 8,000 ft. with midday desert temperature reaching 112F (45C). Takeoff and hovering required far less than maximum power in all cases, including hovering out of ground effect at 8,000 ft. at weights above 15,000 lb. Maximum takeoff weight of the AH-64 is 17,800 lb.

Waypoint entries

The aircraft has an auxiliary power unit, permitting such preflight functions as Doppler navigation waypoint entries and flight control checks to be made before starting engines and turning the rotor. Engines can be started with the rotor engaged or with the rotor held stationary by a rotor brake until the crew is ready for turnup.

The AH-64 uses fixed, wheel-type landing gear rather than skids, which ease ground handling. Raising the collective pitch control lever slightly to about 20% torque and tipping the rotor disk slightly forward with the cyclic stick got the aircraft moving and kept it moving. Slowing and stopping could be accomplished either by reducing power and tilting the rotor disk aft, or simply applying the pedal-activated wheel brakes. Steering was done through the rudder-pedal controlled tail rotor, which provided positive directional control even at minimum power levels.

Hughes Helicopters chief test pilot Stephan Hanvey, with whom I flew on all three flights in the AH-64, demonstrated a rolling takeoff on our initial departure from the Proving Ground's Castle Dome airfield, where the AH-64 test facility is located. I conducted a similar takeoff later in the flight. In each case, only about 35% torque—about a third of the power available—was needed to get the aircraft into the air.

The aircraft lifted off at about 60 kt. on the initial takeoff, while it did not want to stay on the ground above 25 kt. on my subsequent rolling takeoff at a lighter weight. In both cases, a direct vertical takeoff to a hover could have been made with the power that was available.

Low altitudes

The AH-64 is designed to operate primarily at very low nap-of-the-earth and terrain contour flying altitudes. Its control system, which is very responsive for a helicopter, lends itself to this role because it builds pilot confidence quickly. In our case, I first flew the aircraft at altitudes of several hundred to 1,000 ft., gradually coming down to 50–100 ft. as I became more comfortable. In actual combat, the Apache is expected to be operated below treetop level, which is believed to

Much of the AH-64 helicopter's extensive avionics package is mounted in fairings on the lower forward fuselage that are accessible from ground level.

be much less vulnerable and more survivable than higher altitudes when faced with hostile antiaircraft fire.

Handling qualities are among the AH-64's strong points. The sloppiness and slow response typical of most helicopter flight controls are absent in this one, and the rotor and airframe response to control inputs is almost immediate. This made the aircraft easy to hover because fewer control inputs were needed, and it was comfortable to fly at very low altitudes because there was no lag in controls to be concerned with in clearing obstacles.

The AH-64 is equipped with stability augmentation and command augmentation systems that help make the aircraft's flight controls sensitive and responsive. In addition, the helicopter has a 3g to –1g maneuvering envelope, permitting it to be maneuvered briskly and positively around obstructions at low altitude. Army pilots who have evaluated the aircraft said there is no other helicopter in this category with the turning capability of the Apache.

Turn rate

The aircraft does have a high instantaneous turn rate, I found during my flights, although airspeed bled off rapidly during sustained 2.5g turns unless I dropped the nose steeply to sacrifice altitude—when we had it—for speed. Also noteworthy was the apparent capability of the rotor system to endure these maneuvers without any noticeable increase in noise or vibrations.

Ride qualities appeared to be very smooth at all altitudes and most airspeeds during our flights, despite moderate thermal activity over the desert. Above approximately 135 kt. in cruise, a four-per-revolution beat from the rotor system became more obvious with increasing speed, and at an airspeed of 160 kt. the vibrations from this made it difficult to see clearly the television display in the front cockpit.

Climb rates of at least 1,500 fpm were achievable at all altitudes without using maximum power, and within the 0–8,000 ft. altitude where we were operating, climb rates of 1,700–1,800 fpm were possible with maximum climb power. At one point with the aircraft still

Pilot's cockpit—the rear cockpit—in the AH-64 includes electronic attitude direction indicator at center of instrument panel and vertical tape engine and systems displays on left side of panel. The cockpit is roomy and provides excellent forward and sideward visibility.

close to our 16,000-lb. mission gross weight, Hanvey took the controls and pulled the nose up about 20 deg. to enter a climb using just less than intermediate rated power.

As our airspeed dropped, he pushed the nose down to a more level attitude, and our climb rate stabilized at about 1,400 fpm passing through 4,000 ft. We stabilized in a hover at 5,000 ft. for a few moments, before Hanvey pushed the nose down for a steep spiraling descent.

The aircraft was equally agile in sideward and rearward flight. Despite a relatively heavy weight and the over 100F temperatures, the AH-64 was easily controllable in left, right and rearward flight up to its limit of 45 kt. in these directions, and a substantial tail-rotor control margin was evident at all times. Rotor governing was excellent throughout our flights, even when pulling g-forces or when collective pitch was increased or decreased rapidly.

Stable hover

The same easy handling characteristics evident throughout the rest of the flights were apparent when landing. There was no shaking or shuddering during translation from the approach to a hover, and the aircraft could be kept in a stable hover with relatively low power setting and small control applications. By engaging the attitude/hover hold toggle switch on the automatic stabilization equipment panel (rear cockpit only), I was able to take my hands and feet off the controls and let the aircraft hold its own stabilized hover.

The same panel also holds another switch that, when engaged during approach or during nap-of-the-earth flight, places the stabilator attached to the lower aft side of the vertical stabilizer in a 25-deg. trailing edge down position. This position keeps the nose down for better visibility during approaches and slow-speed nap-of-the-earth flight.

Automatic function

Normally the stabilator functions automatically to maintain the AH-64 fuselage in a level flight attitude throughout the flight regime from 30 kt. to the aircraft's maximum speed of 197 kt. In the automatic mode, the stabilator has a range of travel from 5 deg. trailing edge up to 35 deg. trailing edge down. It moves smoothly through this range at rates up to 3.5 deg. per second, and is not noticeable at all in the cockpit except for the movement of a small stabilator position indicator. Because of the stabilator, the pronounced nose-down or nose-up attitudes often seen on many helicopters during accelerations or decelerations are largely eliminated on the AH-64. Even when the nap-of-the-earth/approach mode is engaged on the automatic stabi-

Controls and displays for the Martin Marietta target acquisition/designation system and pilot's night vision system are in the front cockpit of the AH-64, which also has basic engine and flight instruments and a stowable cyclic control stick (center in photo) in case the copilot/gunner must fly aircraft. Optical relay tube at center of panel houses day visual optics and forward-looking infrared displays.

lization equipment panel, the stabilator will program itself to its normal cruise position once airspeed exceeds 80 kt.

All of these features tend to keep the workload at manageable levels on the AH-64, allowing the crew more time to concentrate on accomplishing the aircraft's basic antitank and ground attack missions under night and adverse weather circumstances. This was apparent in a separate flight in the sixth AH-64 prototype, which has most of the elements of the Martin Marietta target acquisition designation system (TADS) and pilot's night vision system (PNVS) that are the core of the AH-64 weapons delivery package. Our 40-min. flight in this aircraft consisted of a simulated night, low-level attack mission with Hanvey occupying the rear cockpit as the pilot while I acted as the copilot/gunner in the front cockpit.

After entering several waypoints into the Singer-Kearfott AN/ASN-128 Doppler navigation set in the front cockpit and getting a short briefing on the TADS sensors normally controlled by the gunner, we started the engines, engaged the rotor and took off toward our first waypoint en route to our simulated target. The Doppler system provided heading indications in both the front and rear cockpits to steer us toward the first waypoint. In the meantime, I used the TADS to observe our progress rather than looking directly outside.

TADS sensors include direct-view optics much like a monocular telescope, forward-looking infrared, multipower television, a laser range-finder/designator, and a laser tracker for spotting and directing weapons toward targets designated by other airborne or ground-based laser designator units. All of these are operated by the copilot/gunner. The pilot has his own infrared night vision system, which the copilot/gunner also can use.

We did not use the Honeywell helmet-mounted monocular sights that will be routinely employed by Army AH-64 crews, but most of the symbology that would normally be viewed through the helmet sight was visible on the optical relay tube that protrudes from the center instrument panel in the front cockpit. This is actually more than one tube and consists of an eyecup at the top for looking into the direct-view optics, and a small video screen beneath the eyecup for viewing television and infrared imagery as well as laser symbology from the TADS and PNVS. Handgrips on each side of the optical relay tube contain all the buttons and switches necessary for the gunner to operate the sensors and the aircraft's weapons systems with minimal movements about the cockpit.

Waypoint selection

We flew out to our first waypoint, a small bridge, at 1,500 ft. As we passed over the bridge, I selected waypoint No. 2 on the Doppler navigation system, and Hanvey began a descending turn to the right as the steering cues in both cockpits swung in that direction to indicate the location of the next waypoint. Distance to the waypoint was indicated on the pilot's electronic attitude director indicator (EADT) in the rear cockpit and on the Doppler control display unit in the front cockpit.

We were nearly at ground level, well below the hilltops, as we passed the second waypoint and I selected waypoint No. 3 on the Doppler system. Hanvey turned the aircraft to follow the Doppler steering cues and began slowing the aircraft for nap-of-the-earth flying in simulated night conditions from that point to the target. With the aircraft only about 20–30 ft. above a dry desert wash, we proceeded on our course at an airspeed of about 30 kt., which Hanvey said was about as fast as a crew would normally operate at this altitude at night or in bad weather.

I was using the TADS infrared system for outside vision during this phase of the flight. The imagery on the optical relay tube was not particularly clear or distinct, although this could be blamed partly on the bright sunlight shining on the unshaded video screen. This problem would not have existed under actual nighttime conditions. Hanvey said this was one of the original TADS units, and in addition to normal wear and deterioration, it had not been updated with all of the improvements introduced during the development program.

In the rear cockpit, Hanvey was using the EADI to provide the attitude and steering information he needed to fly the aircraft to the target. The pilot's cyclic control stick has a four-position switch at the top of the stick that can be used to select the appropriate information for the cruise, transition, hover and bob-up (pop-up for weapon delivery) phases of the mission. With the mode switch in the "cruise" position, the EADI provides:

- A 180-deg. heading scale across the top of the display with actual heading at the top center, sensor pointing direction indicated with a solid triangle, and desired aircraft heading (steering cue) indicated with an inverted "V."
- Digital altitude display at right side center, with an analog altitude strip appearing below 200 ft.
- A solid triangle on the right side indicating rate of climb.
- PNVS gimbal limits (when the sight select is in the PNVS position), represented by a rectangular box at the bottom center of the display.
- Integrated helmet and display sight system field of view, a smaller rectangular box within the PNVS gimbal limits box at the bottom of the screen.
- Diamond-shaped aircraft symbol, horizon line and sideslip display.
- Engine torque digital readout at upper left.
- Absolute airspeed digital readout at center left.
- PNVS line-of-sight reticle, a standard cross-hair arrangement open in the center.

All of this symbology can be displayed on the pilot's helmet-mounted sight so that he has the information he needs to fly the aircraft in night/adverse weather conditions while still maintaining a constant outside scan.

As we approached our third waypoint, from which we intended to simulate an attack on a target, Hanvey began slowing the helicopter and selected the "transition" flight symbology mode on his

EADI. This provided him with the cruise symbology he had selected previously as well as:

- A velocity vector consisting of a heavy line with a solid black dot at the end of it extending from the PNVS reticle to indicate the direction and rate of aircraft movement.
- An acceleration cursor, represented by a small empty circle, to provide the pilot with a predicted magnitude and direction of aircraft movement.

We came to a hover at our third waypoint, hidden behind a hill from our target—a water tower at the Castle Dome airfield we had left a few minutes before. Entering the hover, Hanvey selected the "hover" symbology mode for his EADI, eliminating the horizon line from the display and increasing the sensitivity of the velocity vector and acceleration cursor so that any drift from our hovering position would be indicated immediately and could be corrected.

As we came to a hover, the helicopter raised a cloud of dust that was partially visible in the Flir, but did not obscure our vision significantly. Had this been an actual night mission, which is the role the AH-64 is intended to serve, the dust would not have been a problem. In the daylight, however, such a visible signature could have given away our position.

Hanvey backed the helicopter away from the hill in front of us in order to allow the Hellfire missiles we were going to simulate firing to clear the hill after leaving the launch rails. We had two firing options at that point with the Hellfires—locking on before launch, which would have meant popping up and acquiring the target before firing, then firing and holding high hover until the missile reached the target; or lock-on after launch, which meant firing a missile with the known direction and distance to the target, then popping up in time to designate the target with the later while the missile homed on the reflected beam. The latter method would be suitable for known, stationary targets, while the former would generally be used with moving targets such as tanks or other vehicles.

In addition to the Hellfire missiles, the AH-64 also can carry standard 2.75-in. rocket pods on its wing pylons, and it is equipped with a Hughes 30-mm. Chain Gun turret-mounted under the nose. Both these weapons have shorter ranges than the Hellfire.

On the weapons control panel on my front left instrument panel, I selected the waypoint number entered previously with the coordinates of the water tower. This should have directed the TADS sights and steering cues on our displays toward the tower. It did not, however, because of an error made while programming the Doppler system with the waypoints for our flight.

Rather than take the time to reenter the coordinates in the Doppler system, Hanvey popped the helicopter up to a high hover momentarily while I manually boresighted the target win the TADS and slaved the TADS to the target. I also recorded the target position in the fire control computer by pushing a button on the left handgrip of the optical relay tube. From our know position, the fire control computer used the bearing and range to record the position of the target for later recall. In a combat situation, it would have been possible to use the laser ranger to get a very accurate distance fix, and thus a more accurate position fix on the target. We were about 3,500 meters from the target, a typical firing range for the Hughes TOW wire-guided antitank missile.

After simulating a Hellfire launch from that position, we flew to another position about twice the distance from the target, and one more representation of the range of the Hellfire. As we arrived at that point and turned the helicopter toward the heading indicated on our displays for the target, I switched from the Flir display to the day TV display on the video display (and helmet-mounted display, had I been using one), using the switch on the left handgrip of the optical relay tube for the multipurpose sight system in the front cockpit.

All switches needed to operate the target acquisition designation system and night vision system, choose specific displays for the optical relay tube and helmet-mounted sight, operate laser designation and ranging systems, and fire the AH-64's weapons were located on the left and right handgrips on the sides of the optical relay tube (ORT). Thus, as the helicopter approached the firing positions, workload and time requirements were reduced because all the necessary controls were at our fingertips.

In this case, with the target location and range already stored in the fire control computer, I slipped the three-position TADS display switch on the left handgrip from Flir to day position (DTV) with my left thumb, then moved my thumb from that switch to the weapon select switch just beside it, to select the Hellfire missile mode. Moving my thumb further to the left, to the next switch on the left handgrip, I was able to select the wide-field-of-view position for the TV display until we could pick out the water tower target, then moved the switch to the narrow field of view to increase the size of the water tower image in the display. On the right handgrip, I used my right thumb on a coolie-type switch to position the TV reticle directly over the target.

The same switch is used to steer the other sensors—notably the Flir—to the target. Whichever sight is used, pushing a "slave" button on the right handgrip locks the sight reticle on the target.

After simulating another Hellfire missile launch, we departed the area and headed back toward Castle Dome, conducting some handling demonstrations along the way.

First of 11 production aircraft ordered with Fiscal 1982 defense funds is scheduled to be rolled out from Hughes Helicopters production facilities in Mesa, Ariz., on Feb. 29, 1984.

Flight testing of the AH-64 airframe is about 98% complete, according to Maj. Gen. Edward Browne, the Army's program manager for the Apache. A number of improvement efforts not related to the original development contract are under way, he said. These include:

- Qualification of composite rotors for introduction on production aircraft beginning with the third production lot—probably around aircraft No. 100. The AH-64 will be equipped initially with metal blades. Browne said blade production costs will be about $50,000 less per blade with the composite blades, and the composite blades will have a longer fatigue life than the metal blades.
- Reduction in the number of electronic boxes associated with the TADS/PNVS from six to four, resulting in lower weight penalty and lower volume requirements. The new configuration is undergoing qualification testing, and flight testing will begin early next year.
- Refinements to the TADS sight for improved reliability and boresighting. These stem from problems encountered last year during the Army's second operational test and evaluation of the AH-64.

Harsh conditions

The operational tests took place at the Army's Hunter-Liggett military reservation in central California and included some of the most harsh conditions, in terms of dust and heat, that the aircraft could be expected to encounter in service. The helicopter demonstrated high reliability and maintainability and operational availability during these exercises, according to Army test personnel, despite the absence of all but the most basic maintenance equipment and facilities.

Operational aviators and maintenance personnel who have been exposed to the AH-64 so far have been enthusiastic about the aircraft.

"The brute power, the controllability, the tail-rotor power are the most impressive things initially," Chief Warrant Officer Larry Proper, a former Bell AH-1 Cobra attack helicopter pilot before he was assigned to the Apache program, said.

"One of the biggest things you see when you go into a target engagement is the reduced communications requirements between the cockpits, made possible by the simultaneous cueing on the helmet-mounted displays in both cockpits. It provides pass-off capability between the crewmembers when one sees something the other doesn't."

Hit the switch

"The switchology looks more complicated than it really is," Proper said. "Actually, everything is all set up while still on the ground. Then when we get to the target, we just hit the action switch and everything is set to go.

"The three TADS sensors look time consuming, but we can beat the guy with a single sensor every time.

"Earlier, there was a question about whether all these displays were really needed, but we've found that we use them all, and we have the choice of which to use at any given moment, depending on the circumstances. We can find the targets at night and at longer distances than we could before."

During operational tests of the Apache, Proper said, crews were forced to "go low because the threat was out there. We did at least 50% of our flying at night and didn't have a single blade strike. We could hover in the treetops at night and could see the drift indications in the helmet-mounted displays. With the PNVS, we can see where we're going."

Flexibility is offered by XV-15 tilt-rotor

Robert R. Ropelewski/Arlington, Tex.
January 11, 1982

Flight profile flexibility, low workload and relatively subdued noise and vibration levels of the NASA/Army/Bell Helicopter Textron XV-15 tilt-rotor research aircraft are confirming the feasibility of this concept in the evolution of vertical takeoff and landing technology, Bell Helicopter officials here believe.

Flight testing of two XV-15 prototypes by Bell, Army and National Aeronautics and Space Administration pilots has demonstrated the operational practicality of the tilt-rotor concept—a marriage of helicopter and fixed-wing turboprop technology—without uncovering any serious technical drawbacks to the concept.

Early problems resolved

Technical shortcomings and potential problems of earlier approaches to the tilt-rotor concept appear to have been resolved with the XV-15 approach, military specialists believe. Program officials at Bell, confident of the flight procedures and stable flying qualities of the XV-15, have been conducting a guest pilot program in recent weeks with the No. 1 flight test aircraft.

First NASA/Army/Bell XV-15 tilt-rotor research aircraft hovers during test flight prior to Aviation Week & Space Technology's *evaluation flight.*

This *Aviation Week & Space Technology* editor flew the XV-15 in conjunction with the program. Our flight was made here in late December at Arlington Municipal Airport, where the Bell Helicopter flight research center is located. The flight illuminated several positive aspects of the tilt-rotor concept as it exists in the XV-15, and few negative ones.

Because of the limited fuel capacity of the XV-15, our flight time was limited to about 45 min. This constrained the evaluation to a quick overview of performance and handling qualities, but it was sufficient to underscore several noteworthy characteristics of the XV-15, including:

- Uncomplicated cockpit layout and instrumentation, despite the combined fixed-wing and rotary-wing character of the aircraft—Standard stick and rudder pedals control aircraft attitude, while a collective pitch control type power lever at the left side of each pilot's seat serves as the throttle for all phases of flight and collectively adjusts prop-rotor blade pitch in the helicopter mode. Thumb switches at the top of the power lever allow control of rotor rpm and nacelle/rotor tilt angle—referred to as conversion angle—with the left thumb.
- Flight profile flexibility—At any point from takeoff to landing, the crew can transition easily from the airplane to helicopter mode or vice versa without significant procedural changes or extra movement in the cockpit. Conversions require only thumb movements to adjust rotor rpm and change nacelle angles.
- High stability—Tandem rotor configuration of the XV-15 in the helicopter mode neutralizes torque and thus does not require excessive rudder inputs during hovering, in accelerations or decelerations, or with power changes.
- Low workload—Combination of convenient control arrangement and good basic stability allows the aircraft to be operated by a single crewmember under most circumstances, although normal procedures in the XV-15 flight manual call for both a pilot and a copilot.
- Low noise and vibration levels—Inside the cockpit, these seem to be on a par with conventional fixed-wing turboprop aircraft. Vertical beat normally associated with most helicopters in all phases of flight was noticeably absent. Outside the aircraft, XV-15 noise levels are considerably lower than usual with either helicopters or turboprop aircraft.

XV-15 right engine nacelle and rotor are in the vertical, or 90-deg., position prior to starting the 1,550 shp. Lycoming LTC1K-4K (modified T53) turboshaft engine housed in the nacelle. Wing remains in the horizontal position. Double extension concentric jackscrew, which tilts the rotor and nacelle to any selected position from 0 to 95 deg., is shown here extending from wingtip to its nacelle attach point (arrow).

Tilt-rotor concept

The tilt-rotor designation stems from the fact that one or more prop-rotors are used to provide vertical lift for helicopter mode operations, then can be tilted forward to provide forward thrust while the lift function is off-loaded to a fixed wing. In the case of the XV-15, two three-blade, 25-ft.-dia. prop-rotors provide the vertical lift and forward thrust. The rotors are driven by two 1,550-shp. Avco Lycoming LTC1K-4K turboshaft engines mounted in the wingtips.

The LTC1K-4K is a T53 engine uprated and modified for both vertical and horizontal operation. The entire engine, nacelle and prop-rotor package is rotated at the wingtip during conversions from the airplane mode to the helicopter mode and back. The need to remember which mode the aircraft is in at any given moment is the real challenge in flying the XV-15.

Conventional rudders and elevators of the XV-15 have full range of travel throughout the flight envelope of the aircraft, although they are ineffective at extremely slow speeds and in hover. Collective and cyclic movements of the XV-15's two 25-ft.-dia. rotors control the aircraft in hover and at slow speeds while in the helicopter mode, but are gradually washed out as the nacelles rotate forward and/or as airspeed increases.

Concept

Bell experimental test pilot Dorman Cannon occupied the right seat of the aircraft for our flight and demonstrated the start procedure for the engines. These were accomplished with the nacelles in the vertical or 90-deg. position, with rotors parallel to the ground.

Although there was a mild rocking as the rotors began to turn with the engines, the rocking was negligible compared with the response of many helicopters during rotor acceleration.

Once the engines and rotors were started and operating at flight idle, small locking pins on the throttles were dropped into place, linking the throttles to the collective pitch/power control lever beside each pilot's seat. The engine controls and throttles are located at the far aft end of the center pedestal and are not used at all except for engine start and shutdown. The throttles are advanced or retarded mechanically as the power levers are raised or lowered. At 98% rotor rpm, governors were engaged to maintain the selected rotor speed. Throughout the rest of the flight—in both the helicopter and airplane modes—the governors kept rotor rpm locked within a percentage point or two of the selected speed, no matter what the power setting.

Taxiing was initiated by releasing the parking brake and tilting the nacelles about 5 deg. forward from the vertical, or 90-deg. position, using the "coolie hat" beep switch at the top of the power lever. Taxi speed thereafter could be controlled with the left thumb by tilting the nacelles forward or aft as necessary.

The nacelles have an aft conversion limit of 95 deg., or 5 deg. past the vertical, and tilting them to that position was very effective in slowing, stopping and even backing up the aircraft. Use of the wheel brakes was unnecessary except during turns.

Cannon advised carrying a small amount of power while taxiing in order to give the rotor rpm governors a modest operating margin. This made the aircraft light on its wheels and resulted in a mild and somewhat annoying lateral rocking motion as we taxied toward the runway. Test pilots who have been flying the aircraft regularly apparently have become accustomed to this.

Flexibility of the XV-15 offers a range of options to the pilot in all phases of flight, beginning with the takeoff. At gross weights up to 13,300 lb., the aircraft can take off vertically, then accelerate forward in either the helicopter mode or in a conversion mode with the nacelles slowly translating forward to the 0 deg. position. Above 13,300 lb. and up to a maximum gross weight of 15,000 lb., a rolling takeoff is required to become airborne.

Instrument panel in the XV-15 is relatively conventional, and the helicopter/airplane conversion capabilities of the aircraft have not necessitated many new instruments. Exceptions include the dual-needle nacelle position indicator and the mast torque indicators above the glare shield , which measure net torque at the rotors. The torque indicators are the primary power references.

This can be accomplished by tilting the nacelles forward to any angle from 90 to 60 deg., then pulling the power lever to command sufficient power to accelerate the aircraft on the runway until it reaches flying speed. An interlock connected to the landing gear struts prevents the nacelles from being rotated to any less than 60 deg. while the aircraft is on the ground because of the danger of the 25-ft.-dia. rotors contacting the runway.

The aircraft weighed just under 13,000 lb. as we took the runway, and Cannon elected to demonstrate a vertical takeoff initially. With the nacelles at the 90-deg. position—helicopter mode—and rotor rpm at 98%, he lifted the aircraft into a hover at an altitude of about 30 ft. After conducting a power check, he turned the controls over to me.

The air was extremely smooth during our early morning flight and the aircraft remained motionless over the runway after Cannon released the controls. It remained that way for several seconds before it was necessary to pull upward slightly on the power lever to arrest a slow vertical descent.

Although the XV-15 cockpit is equipped with conventional engine instruments and torque gauges, primary power reference is a set of two large rotor mast torque meters mounted on top of the instrument panel glare shield. These were installed in conjunction with the research testing of the aircraft, according to NASA officials, and future production aircraft would probably use standard engine torque instruments for power reference.

Direct power indication

The mast torque meters provide readouts of net torque on the two rotor masts—the most direct indication of power being applied. Mast torque is limited to 130,000 in./lb., although the Lycoming engines are capable of producing 160,000 in./lb., Cannon said.

With the cool temperatures prevailing, and at our density altitude close to sea level, a mast torque of about 100,000 in./lb. was sufficient to maintain our 30-ft. hover. Hovering turns over a spot on the runway, sideward flight at slow speeds and rearward flight were all accomplished easily. Left to itself, the XV-15 displayed excellent stability in all axes and demanded little in the way of control inputs.

The aircraft is equipped with a stability and control augmentation system, which was left on for the time being. Later in the flight we turned the system off, with no significant increase in workload necessary to control the aircraft.

A rotor rpm of 98%—100% rpm is equivalent to 601 actual rpm—is used for the hover mode in the XV-15, and the nacelles and rotors cannot be rotated from the 0 deg. (full forward) position until rpm has been set to at least 98%. Design rotor rpm in the airplane mode was 76%, but a speed of 86% has been used so far in flight tests because of vibratory loads that were encountered when operating at the lower rotor speed.

Flight limitations

Rearward and sideward flight limitation for the XV-15 was predicted to be about 35 kt. The aircraft has been cleared to this speed in sideward flight, but so far has only been demonstrated to 25 kt. in rearward flight.

While the aircraft had to be banked several degrees in either direction for sideward flight, it was possible to keep the longitudinal axis of the vehicle relatively level because of the facility of tilting the nacelles to accelerate fore and aft.

Cannon took control again and tilted the nacelles to the 95-deg. position—5 deg. aft from vertical. This resulted in a strong rearward acceleration of the aircraft. Without changing power, he beeped the nacelles forward to the 70-deg. position, and the aircraft quickly slowed, stopped its rearward flight and accelerated forward. As the forward acceleration began, Cannon raised the landing gear, which has a 90-kt. extension limit. Flaps were raised from 40 deg. to 20 deg. as airspeed passed through 60 kt., and then were fully retracted as acceleration continued.

Cannon continued to rotate the nacelles to 0 deg. He then applied the pylon lock by moving his thumb just above the nacelle conversion switch to the pylon lock switch, also located on the head of the power lever. The pylon lock applies a positive downward pressure on each nacelle to hold it in the 0 deg. position for high-speed cruise. It also eliminates a vibration mode detected in the nacelles during earlier flight tests.

Full conversion

With the aircraft now fully converted to the airplane mode, rotor rpm was reduced to 86% using the thumb switch at the lower left corner of the power lever head. A high-frequency vibration accompanied by an increase in noise from the rotors was noted as aircraft speed increased with the rotors still at 98% rpm, but these phenomena disappeared when rotor rpm was reduced to 86%.

The XV-15 was now essentially a conventional fixed-wing turboprop aircraft. As the nacelles had rotated to the full forward position and airspeed had increased, the helicopter mode control inputs to the rotors had been gradually washed out so that now the rotors were straightforward turboprops.

In the helicopter mode, power lever motion changes rotor blade collective pitch directly to provide a more responsive and direct control of vertical thrust. This collective pitch response to power lever motion is mechanically phased out as nacelle angle decreases, so that in the airplane mode the power lever controls only power and not blade pitch.

Cyclic pitch and roll inputs to the rotors from the control stick are also phased out mechanically at reduced nacelle angles, while yaw inputs to the rotors from the rudder pedals are phased out as airspeed increases and the twin rudders of the XV-15 become effective.

The XV-15 quickly accelerated to 230 kt., which test crews have been using as normal cruise speed for the aircraft. Cannon executed a series of low-altitude maneuvers including tight turns left and right at 60 deg. angle of bank, a high-speed pass along the runway, and a reconversion to the helicopter mode and subsequent steep approach to a taxiway with nacelles tilted aft to 95 deg.

We taxied onto the runway once again, and Cannon executed a short-field takeoff using simulated single-engine power. With the aircraft lined up on the runway, flaps set at 40 deg. and nacelles tilted to their 60-deg. position—the lowest allowable with the aircraft on the ground—Cannon held the brakes and brought the power up to 65,000 in./lb. of torque on each rotor mast. This was half the allowable torque, and simulated operation with a single engine. Transmissions are interconnected through the wing, and in case of a single-engine failure, the remaining engine would continue to drive both rotors.

Simulating this condition in the XV-15, Cannon released the brakes and the aircraft accelerated quickly, reaching 50 kt. in four or five seconds and lifting off after a roll of just a few hundred feet. Gear and flaps were raised and nacelles were rotated forward to the 0 deg. position for a climbout in the airplane mode.

Wide corridor

The conversion corridor—the difference between wing stall speed and rotor overstress speed at any given nacelle angle—is comfortably wide in the XV-15. As a result, although the airspeed/nacelle angle relationship must be observed throughout the flight, an inordinate amount of attention is not necessary. At a nacelle angle of 60 deg. and a gross weight of 13,000 lb., for example, the aircraft has a wing stall speed of 54 kt. at the low speed end of the conversion corridor and a rotor stress limit of 160 kt. on the high speed side of the corridor. Below 54 kt., the rotor must be in the full helicopter mode of 85–95 deg. to avoid a stall, while above 160 kt. the rotors must be in the 0 deg. airplane mode to avoid overstressing.

Normal rate of nacelle conversion is 7.8 deg. per sec., so the nacelles can translate through their full 95 deg. of rotation in 12–13 sec. They also can be stopped at any position between the extremes by releasing the nacelle beep trim position switch on the power lever head. Both the pilot and copilot have a small dual-needle nacelle position indicator on their instrument panel to show at a glance the position of the rotors and pylons. The rotors and pylons also are easily visible out the side windows of the XV-15 cockpit.

We had established a cruise climb with a speed of about 200 kt. after departing the airport, using a power setting of 110,000 in./lb. on each rotor to give us a climb rate of about 2,000 fpm. Cannon noted that climb rates of over 3,000 fpm are possible at a climb speed of 160 kt. using maximum available power of 130,000 in./lb.

I leveled the aircraft at 6,000 ft. and left the power at 100,000 in./lb. mast torque on each rotor. The XV-15 stabilized at a speed of 210 kt. indicated airspeed (220 kt. calibrated airspeed). The aircraft was smooth and relatively quiet under those circumstances. Flight controls exhibited fixed-wing responsiveness in. a series of 90-deg. turns left and right, and the nose remained stable on the horizon, making it easy to maintain our altitude.

Although there is a small amount of differential collective pitch control programed into the rotors even in the airplane mode to counter adverse yaw during turns, it was still necessary to pay close attention to the slip indicator to counter adverse yaw with a half inch to one inch of rudder pedal in the appropriate direction.

At Cannon's suggestion, I reduced mast torque to 15,000 in./lb. on each prop-rotor to slow the aircraft for a stall. Flaps were extended 20 deg. at 150 kt. and left in that configuration as the aircraft slowed.

Mild buffeting could be felt in the controls at about 110 kt., and the aircraft stalled at approximately 102 kt. indicated airspeed. The stall was straightforward, with no tendency for either wing to drop. Flaperons—combination flap/ailerons—remained effective, as did the elevator, and the nose dropped immediately when the stick was released. With the application of about 70,000–80,000 in./lb. of torque on each rotor, the aircraft recovered quickly.

While the aircraft was still under 150 kt., I increased prop-rotor speed from 86% to 98% using the thumb switch on the power lever. At 98%, the pylon lock was released and I beeped back the nacelles to an intermediate position of about 60 deg. With power remaining at approximately 80,000 in./lb., the airspeed stabilized at about 100 kt. Navy evaluators note this would probably be a good loiter configuration and speed for maritime surveillance applications.

I brought the nacelles back further to about 70 deg. without changing power, and the indicated airspeed slowed to 90 kt. This was an effective demonstration of the greater efficiency of a fixed-wing aircraft over a helicopter in these circumstances. For the same power setting and fuel consumption, the speed of the XV-15 in the helicopter mode was about half what it was in the airplane mode. Bell program officials claim the tilt-rotor concept is about 40% more efficient than an equivalent size conventional helicopter in a cruise situation.

This is because of the inherent speed limitations of helicopters due to the increasing aerodynamic inefficiency of a rotor flying edgewise at high speeds, they said. Aerodynamic limitations of helicopter rotors become more acute as speed increases and altitude increases.

High-frequency vibration

I beeped the pylons full forward to the 0 deg. position at that point, accelerating again to a normal airplane mode. As airspeed increased, the same high-frequency vibration and noise occurred that had been noted earlier during a conversion, before rotor rpm was reduced from 98% to 86%. I beeped back the rpm to 86% and the vibration and noise disappeared. Cannon disengaged the No. 2 throttle from the power lever and brought it to flight idle to simulate an engine loss.

The interconnect between the two nacelles continued to drive both proprotors at the same speed, and there was no asymmetric yaw or any other noticeable changes in the flight control system. Airspeed began to decay slowly, and I raised the power level to 65,000 in./lb. of torque on each rotor—the maximum allowable with both driven

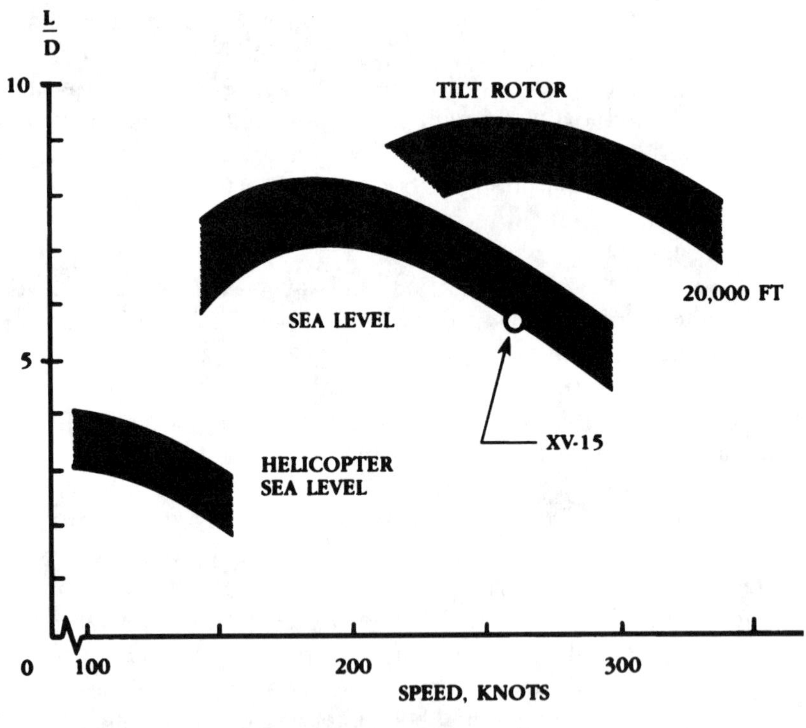

Relative efficiency of tilt-rotor aircraft is shown in this chart comparing the lift-to-drag ratio envelope of typical helicopters at sea level with tilt-rotor aircraft at sea level and 20,000 ft.

by a single engine. The aircraft stabilized on speed and altitude after losing only a few knots.

Cannon brought the No. 2 engine back on line again, and we began a descending turn toward the airport. After making a high-speed low pass across the field, we pulled up and entered a left downwind leg for the runway.

As we turned onto the downwind leg, I beeped the rotor rpm to 98% from 86% and began slowing for the approach. Flaps were extended to 20 deg. as our airspeed dropped below 150 kt., and the pylons were beeped aft to their 60-deg. position, further slowing the aircraft to about 110 kt. Flaps were extended to 40 deg., the maximum deflection used even though a 75-deg. setting is available.

Decreased speed

After the turn onto base leg, airspeed dropped below 90 kt. and the landing gear was extended. Pylons were slowly rotated further aft from 60 deg. to 80 deg. before the turn was begun onto final. With some excess airspeed still to be lost, Cannon rotated the nacelles aft to 95 deg. for a few moments for some braking action, then beeped them back to 90 deg. for the remainder of a steep approach and gentle touchdown using a slight pull on the power lever to cushion the landing.

Taking the aircraft back up into a hover, I beeped the pylons forward to 70 deg. while maintaining a level fuselage attitude. The aircraft accelerated forward, giving a slight shudder as it passed through translational lift. A small amount of nose-up trim was needed as the XV-15 continued to accelerate to about 130 kt., and we turned downwind for our final approach to the airport.

It was possible to fly the aircraft down to a 30-ft. hover with almost no power adjustments after turning off the downwind leg, using only nacelle conversion angle to control the airspeed and rate of descent of the XV-15.

As on earlier touchdowns and in other phases of the flight, only limited rudder displacements were necessary, and power adjustments did not necessitate a chain of control responses in all other axes.

Turbulence response

Although other pilots who have flown the XV-15 have said it displays excellent ride characteristics in turbulence, the smooth quality of the air on the morning of our flight precluded any observation of turbulence response. One unusual trait noted in the XV-15 in gusty air has been a longitudinal acceleration and deceleration phenomenon re-

Box at top of power control lever in XV-15 holds switches for converting from helicopter to airplane mode of flight and back again. Beep switch at lower left corner of box increases or decreases rotor rpm as necessary for any given mode of flight. "Coolie hat" beep switch at lower right of box rotates engine nacelles and rotors from the full forward, or 0-deg. position (airplane mode) to the vertical, or helicopter, mode. Switch at upper right locks pylons in full forward position for high-speed flight in airplane mode. Switch at upper left disengages rotor rpm governors while button at extreme lower right is part of flight test package controlling flight data recorder.

ferred to by some pilots as "chugging." Instead of the vertical accelerations felt in most aircraft in gusty conditions, this is a fore and aft surging effect, which program officials say is caused by transient changes in blade angle of attack. These changes produce large propeller thrust changes which, in moderate turbulence, generate fore and aft motions that have at times been uncomfortable to the aircraft's pilots. Modifications to the rpm governor and stability and control augmentation system will be able to resolve this problem, program officials believe.

Workload for pilot reduced in SH-60B

Donald E. Fink/Stratford, Conn.
December 1, 1980

Navy/Sikorsky SH-60B Seahawk helicopter, the airborne portion of Navy's Lamps Mk. 3 antisubmarine warfare system, offers a combination of automation, performance and crisp handling qualities that should reduce the workload on pilots flying operational missions from their small shipboard platforms, especially under bad weather conditions.

Performance of the SH-60B Seahawk was evaluated here recently by this *Aviation Week & Space Technology* pilot during a 1-hr. flight in the No. 2 prototype. W. Richard Faull, Seahawk project pilot, was in the right seat.

Under the Lamps (light airborne multipurpose system) concept, SH-60Bs will be carried on hangar decks aft of the superstructure on destroyer, frigate and cruiser class ships (AW&ST June 30, p. 47). The Lamps helicopters will be fitted with a variety of antisubmarine warfare (ASW) sensors, sonobuoys and two air-launched Mk. 46 torpedoes.

The helicopters will operate under realtime data link control of a combat information center on their ships, enabling them to conduct over-the-horizon ASW and antiship surveillance and targeting (ASST) missions.

IBM's Federal Systems Div. was selected by the Navy as prime systems integration contractor for Lamps to assure that the critical link between the airborne sensor packages and the combat information center functions properly (AW&ST Feb. 12, 1979, p. 52). The other major Lamps contractors are Sikorsky for the SH-60B, General Electric for the T700-GE-401 turboshaft engines and DAF Indal, Ltd., for the ship deck-mounted recovery, assist, securing and traversing (RAST) system.

The requirement to operate from the limited space provided by the small ship hangars and helipads dictated several design modifications on the Seahawk, which is a derivative of the UH-60A utility helicopter Sikorsky developed for the U.S. Army.

Major airframe change involved addition of an electric folding system for the main rotor blades and the tail pylon that makes it possible to reduce the Seahawk to a stowed size that can be accommodated on the small decks. Folding the tail pylon necessitated relocating the tail wheel forward of the hinge line on the aft fuselage.

A special probe system also was incorporated with the fuselage centerline cargo winch to function with the RAST system on the carrier ships. The deck system includes a winched cable to guide the helicopter to touchdown, a probe lock that secures the helicopter to a heavy sled on the deck and tracks to facilitate deck handling of the helicopter once it is attached to the sled.

Other external changes on the Seahawk include the addition of various radar and data link antenna domes and stores stations associated with the SH-60B's primary ASW and ASST missions (AW&ST Sept. 3, 1979, p. 21).

The *Aviation Week & Space Technology* evaluation flight originated at Sikorsky's main production facility here and included hover and pattern work from the test center heliport, airwork over Long Island Sound and a Tacan approach to Runway 23 at Calverton Airport on Long Island before returning to Stratford.

A thorough preflight inspection had been completed by a Sikorsky ground crew, so Faull and I had to conduct only a brief walkaround inspection before strapping into the cockpit for the pre-engine start briefing. Since the No. 2 prototype is one of five flight test articles, a flight test engineer accompanied us to handle the flight data recording equipment in the aft cabin.

On operational missions, the Seahawk will carry a three-man crew consisting of a pilot, an air tactical officer who also can function as the pilot, and a sensor operator who runs a control and data display station in the aft cabin.

Maximum gross weight at engine start was 20,500 lb., with the three-man crew and full complement of test equipment. External stores included a dummy magnetic anomaly detector mounted on a pylon on the aft right side of the fuselage.

Mission takeoff gross weight for the Seahawk is 20,244 lb. for ASW and ASST operations. Maximum takeoff gross weight is 21,844 lb. Engine start on the Seahawk is fully automatic and designed to get the helicopter airborne in a very short time after it is positioned on the aft deck helipad. Faull said he has demonstrated the capability to start the engines and take off in less than 1 min.

The start sequence was initiated by starting the auxiliary power generator. For operational missions, the shipboard high-pressure air source also can be used. Engine start buttons can be pushed one at a time, or simultaneously if a rapid start is desired, as the throttles are moved from the stop-cock position to ground idle to get engine light.

The rotor brake was left locked until both engines were up to ground idle. This eliminates slow rotor rotations with drooped blades, Faull said. During deck operations in heavy sea states, a slow moving blade could be a hazard to deck personnel or could impact part of the helicopter airframe. With the RAST system, the SH-60B is cleared to operate from a deck that is rolling through angles of 28 deg., pitching 5 deg. and heaving up to 15 ft. vertically.

Faull started the engines separately, and when both were stabilized at ground idle, he released the rotor brake and completed hydraulic flight control and leak detection tests. The throttles then were advanced to the 100% position. At that point, they dropped into a detent on the throttle quadrant and did not have to be touched until engine shutdown at the end of the flight.

Exhaust gas temperature on both engines peaked at about 710C before stabilizing at 550C at ground idle. Hot start limit on the T700 engine as installed in the Seahawk is 850C. So far in the test program, that limit has never been encountered, Faull said.

On receipt of clearance to taxi, Faull turned the helicopter over to me and recommended a collective setting of between 20–25% torque for taxi. The flat rotor torque range is 18–19%. Following brake release, the Seahawk proved to be quite responsive and easy to taxi. A slight roll input with the cyclic was used during turns to compensate for the Seahawk's angled tail rotor. Normal operating limits are 258 rpm for the main rotor and 1,154 rpm for the tail rotor.

After taxiing to the Sikorsky test center helipad, I lifted the Seahawk into a hover. Anticipating heavy control forces, I tended to overcontrol the helicopter for the first few seconds until Faull coached me into using a light touch and letting the stability augmentation system fly the machine.

Despite its size, the Seahawk gives the impression of being a small, highly maneuverable machine. Fuselage length of the SH-60B is 50.7 ft., overall length with the main rotor turning is 64.8 ft., height is 17.1 ft. and fuselage width is 7.7 ft.

After I had completed a sequence of hover performance evaluations, including vertical takeoffs and landings, sideward flight and liftoffs and touchdowns with my feet off the antitorque pedals, Faull took command of the helicopter and flew a traffic pattern to demonstrate how the automatic flight control system functions on a full approach without torque pedal inputs. The system held the nose steady through the initial approach, the flare to hover and touchdown.

Three separate packages compose the Seahawk's automatic flight control system—the variable stabilator, a dual stability augmentation package and the autopilot, Faull said. The stabilator automatically rotates—its limits are +8 deg. trailing edge up and –40 deg. trailing edge down—as the nose is pitched up or down. This enhances the pitch handling qualities, minimizes pitch excursions and improves pilot visibility when the helicopter is flared by reducing the pitch angle.

The stability augmentation system consists of dual units—one analog and one digital—that provide damping in all three axes. Using analog and digital units results in a combined system that offers both reliability and economy, Faull said. The stability system performs an inner loop function, which the pilot is not aware of until the system is turned off or fails in flight.

The outer loop function, which the pilot is involved in, is performed by the autopilot, which offers heading hold, altitude hold, airspeed hold, attitude retention and longitudinal stability.

The heading hold function has accuracy limits of ±3 deg., but experience with the system so far indicates the actual performance is closer to ±1 deg., Faull said. The airspeed hold function uses short-term correction information from an attitude sensor and long-term correction information from accelerometers. Its accuracy is in the ±2–3 kt. range.

Dual altimeters—one barometric and one radar—are used to provide data for the altitude hold function, and the attitude retention uses yaw rate and vertical rate gyros. The autopilot also has three coupled functions that permit SH-60B pilots to select automatic approach, auto hover and automatic departure.

After I had flown the Seahawk out over Long Island Sound for airwork evaluations, Faull set up an automatic approach to a hover altitude of 40 ft. above the water. We were cruising at an altitude of 1,500 ft., and the autopilot established the helicopter in a 360-fpm descent. It held this rate until we reached a radar altitude of 200 ft. above the water. At this point, the autopilot commanded a 1 kt./sec. deceleration and slowed the rate of descent to 120 fpm.

The combined deceleration and descent rates were held until the helicopter was brought smoothly to a hover at 40 ft. A higher or lower hover altitude can be selected, Faull said, but 40 ft. is what is used routinely.

From the hover, Faull demonstrated an automatic departure by selecting that mode on the autopilot console. This established the helicopter on a 500-fpm climb profile with an acceleration of 2 kt./sec. The automatic departure sequence concluded with the helicopter stabilized at a speed of 100 kt. and an altitude of 500 ft. "At any altitude below 500 ft. and any airspeed below 100 kt., the pilot can take over manually by pushing the collective trim release to stop the climb or the cyclic trim release to stop the acceleration," Faull said.

During departures, whether automatic or flown manually, the Seahawk accelerates smoothly through the translational lift regime.

On final approach—especially in the automatic mode—the helicopter spends about 10–15 sec. in the translational lift regime, and a definite four-beats-per-revolution vibration is felt. The vibration levels vary from prototype to prototype, and the No. 2 machine I flew has the highest level.

The vibration poses no control problem and would not distract an experienced operational pilot, Faull said. The UH-60A Black Hawk has a similar characteristic, but Army pilots, who are trained to fly assault approaches, spend far less time in the translational lift regime, and encounter the vibration for shorter periods of time.

During high-speed runs over the Sound—a maximum speed of 155 kt. was reached—the Seahawk exhibited very smooth flight characteristics. Vibration and noise levels were lower than those I had experienced in several current generation executive helicopters.

The No. 2 prototype was still operating under maneuvering restrictions because structural testing of the Seahawk airframe had not been completed, but Faull flew the helicopter through a series of tight turns that demonstrated it has excellent high-speed handling qualities.

At another point during the high-speed airwork, Faull took control of the helicopter and executed a 2g pullup followed by a negative-g-force pushover. Torque pedal and cyclic control inputs during the pushover showed the Seahawk retained crisp lateral and roll control throughout the maneuver.

After I had completed the Tacan approach to Calverton and was flying the return trip across Long Island Sound to Stratford, Faull shut off all three elements of the automatic flight control system. The helicopter remained controllable, but there was a significant increase in the pilot workload required to maintain constant altitude, airspeed and heading.

Faull then switched on each element of the automatic flight control system in succession, which gave a graphic demonstration of the function performed by each loop of the system.

With the automatic flight control system fully restored and the helicopter stabilized at a cruise speed of 100 kt., Faull chopped the throttle on the No. 2 engine to idle to simulate an engine failure. A stop bracket is fixed on each throttle to prevent an inadvertent inflight shutdown of an engine during such demonstrations.

I sensed no yaw as the No. 2 engine spooled down, and no manual trim was needed to maintain heading. The No. 2 engine then was brought back to 100% and the helicopter was accelerated to V_H, which is 155 kt. IAS with an intermediate rated power setting of 106%

torque. Faull chopped the No. 2 engine a second time, and again there was minimal effect other than a bleed-off in airspeed.

Autorotational performance was not included in the evaluation flight because the prototypes were still restricted to simulations with full power maintained throughout a power recovery. Power-off autorotations to touchdown will be demonstrated later in the Seahawk flight test program as they were in the UH-60A Black Hawk program, Faull said.

SH-60B Seahawk Specifications

	ASW	ASST
Empty weight	14,019 lb.	14,019 lb.
Maximum takeoff gross weight	21,884 lb.	21,844 lb.
Mission gross weight	20,244 lb.	18,353 lb.
Hover weight out of ground effect	21,844 lb.	21,844 lb.
Dash speed		
(sea level standard day)	145 kt.	145 kt.
(sea level 90F)	147 kt.	147 kt.
Exceeds required mission station loiter time by . . .	52 min.	45 min.
Exceeds desired mission station loiter time by . . .	17 min.	15 min.
One engine inoperative rate of climb	600 fpm.	600 fpm.
Two engine vertical rate of climb (takeoff gross weight, 100%, sea level and 90F)	980 fpm.	No requirement

Dimensions

Overall length (rotors turning)	64.8 ft.
Fuselage length	50.7 ft.
Overall height	17.1 ft.
Fuselage width	7.7 ft.
Main rotor	
Diameter	53.6 ft.
Number of blades	4
Material Composite—titanium spar—Nomex honeycomb filler and fiberglass skin	
Tail rotor diameter	11.0 ft.
Wheel base	29.0 ft.

ASW—antisubmarine warfare

ASST—antiship surveillance and targeting

Agusta A129 marks European entry into 1990s combat helicopter design

Robert R. Ropelewski/Cascina Costa, Italy
June 8, 1987

Agusta Group has completed tests aimed at exploring the full flight envelope of its new A129 Mangusta light attack helicopter and now is focusing on refining the A129's integrated mission equipment package in preparation for initial production deliveries to the Italian army early next year.

The introduction of the A129 into the Italian army's helicopter fleet will mark the entry into service of Europe's first helicopter designed and developed from the ground up for combat operations in the threat-intensive battlefields envisioned for the 1990s. At the same time, the level of technology and systems integration being applied to the A129 mark it as a forerunner and potentially significant competitor to the LHX family of light helicopters still in the definition phase for the U.S. Army.

Two Agusta A129 Mangusta light attack helicopter prototypes hover over water during flight testing in northern Italy. Agusta is developing the 3,800-kg./8,400-lb. aircraft to Italian army specifications, but is marketing it internationally as well. First production deliveries to the Italian army are scheduled to begin early next year.

This *Aviation Week & Space Technology* pilot flew two orientation flights in the A129 recently for a preview of the capabilities that will be available to the aircraft's eventual users. Although neither of the two aircraft flown was equipped with the full A129 mission equipment package, the aircraft demonstrated significant advances toward the levels of agility, performance and handling qualities that will be necessary on the high-threat battlefields of the future. Both of the *Aviation Week* flights were flown from the Agusta Group's experimental flight test facility here at Cascina Costa, northwest of Milan.

Evolutionary design

Design of the A129 evolved from a series of Italian and European multinational studies conducted during the 1970s. These led to detailed design in the early 1980s and the start of flight testing in September, 1983 (AW&ST Sept. 19, 1983, p. 29).

The operational requirement for the aircraft is to perform antiarmor and area attack missions day and night, under instrument and light icing conditions, in the North Atlantic Treaty Organization theater during the 1990s. Primary armament consists of eight Hughes Aircraft Co. TOW wire-guided antitank missiles, but the aircraft also is configured to carry an array of nonprecision area weapons such as 2.75-in. rockets and turret-mounted 50-cal. and 12.7-mm. machine guns.

In addition, provisions are incorporated for the firing of Rockwell International Hellfire laser-guided antitank missiles, other future antiarmor weapons, and air-to-air missiles including the Ford Aerospace/Raytheon AIM-9 Sidewinder, General Dynamics Stinger and Matra Mica.

The A129 is relatively uncramped inside. Its cockpit arrangement provides for tandem seating of a two-man crew, including a copilot/gunner in the forward station and a pilot in the rear position. Outside visibility is excellent from both seats, with the pilot's station raised high enough above the copilot/gunner's that there is no noticeable obstruction of forward visibility from the rear cockpit.

Communications, navigation, hydraulic, fuel and electrical systems, as well as flight controls, engines, avionics and weapons systems, are all controlled and monitored through the A129's integrated multiplex system. The system is built around a MIL-STD-1553B multiplex data bus and uses two central computers to provide redundant data processing capabilities.

Harris Corp. is responsible for overall integration of A129 systems. The integrated system defined for initial production aircraft uses only 75% of the processing capacity available, leaving room for the addition of other capabilities later.

A control display unit (CDU) with alphanumeric keyboard is mounted on the left console in each cockpit to allow the crew the system functions they desire. A multifunction cathode ray tube display on the left forward instrument panel in each cockpit provides a wide variety of display options, including checklists, flight plan and navigation data, horizontal situation indications, aircraft performance computations, status of weapon systems and caution and warning displays.

One of the objectives of the integrated systems approach is to reduce crew workload during low-altitude combat operations at night and in adverse weather. However, there were occasions during my two flights in the A129 when it seemed that too much time had to be spent with the CDU keyboard, and it would have been better to have other means such as switches on the cyclic stick or the collective lever.

Agusta chief test pilot Raffaele Longobardi, with whom I flew both flights in the A129, acknowledged the possibility of using switches on the cyclic and collective to select some functions. However, he said no decision will be made until after the Honeywell integrated helmet and display sight system has been fully integrated in the A129 and the requirements have been determined for IHADSS controls on the cyclic and collective.

The first of two *Aviation Week* flights in the A129 was flown in the No. 3 prototype. This aircraft originally was intended to be used as a ground test article, but was adapted as a demonstrator aircraft at the request of the Italian army. Because of this, it is not fully representative of a production aircraft. The front cockpit, from which I flew the first of the two A129 flights, had none of the A129 mission equipment package installed, nor did it have the multifunction CRT display that will be standard in production series aircraft. In addition, this particular prototype is equipped with a mechanical flight control system in the front cockpit rather than the fly-by-wire system planned for production aircraft. Performance characteristics, however, were representative of the production A129, according to Longobardi.

TOW missiles

The aircraft was equipped with four dummy TOW missiles on each stub wing, ballast for the missing TOW sight unit, other uninstalled mission equipment and several hundred pounds of test equipment. A full load of 600 kg. (1,320 lb.) of fuel and the weight of the crew brought the takeoff weight of the aircraft close to the A129's maximum gross weight of 3,900 kg. (8,580 lb.). Mangusta prototypes are powered by two Rolls-Royce Gem 2 Mk. 1004 turboshaft engines with a takeoff/30-min. power rating of 975 shp. and a 20-sec. emergency rating of 1,035 shp. each.

A production aircraft will be equipped with Mk. 1004D engines generating 1,050 shp. for takeoff. The engines can be started using either external ground power or electrical power from on-board batteries. In this case, we used external ground power to start the first engine. The rotor began turning and accelerating with engine start, and once both engines were running, Longobardi advanced the power levers to bring rotor speed up to 100% before liftoff.

Composite blades

The A129 uses an Agusta-designed, fully articulated main rotor with four composite blades attached by composite grips to a steel hub. Four elastomeric bearings, one for each blade, absorb lead-lag, flapping and pitch change loads on the blades. As demonstrated on this flight, the design provides exceptional vibration damping and control response characteristics.

Rotor turn-up with engine start was notably smooth, with none of the rocking normally associated with this action in most helicopters. Low vibration levels were routine throughout this and the following flight with only one exception, which was observed later in the first flight. With the takeoff checklist completed, Longobardi lifted the helicopter into a hover and air taxied to an area away from the main parking ramp, where I took the controls. The Cascina Costa facility is at an elevation of 700 ft., temperature was a cool 5C, and winds were light and variable at the start of the flight.

Without the stability and control augmentation and automatic flight control systems that will be standard in the production aircraft, the mechanical flight controls in the front cockpit were extremely sensitive. The result was an almost jittery response to control inputs that took a few minutes to become accustomed to.

The fly-by-wire flight control system that will be used in the front cockpit of production A129s is being tested on an "iron bird" hydraulic test rig. All of the components of the control system will be installed by the end of this year in the No. 5 prototype aircraft, which will be equipped with the complete integrated multiplex system planned for the A129. Flight testing of the fly-by-wire cyclic controls will not begin until early next year, however.

With the cool temperatures that prevailed, only about 80% torque was required to maintain a hover in ground effect at near-maximum

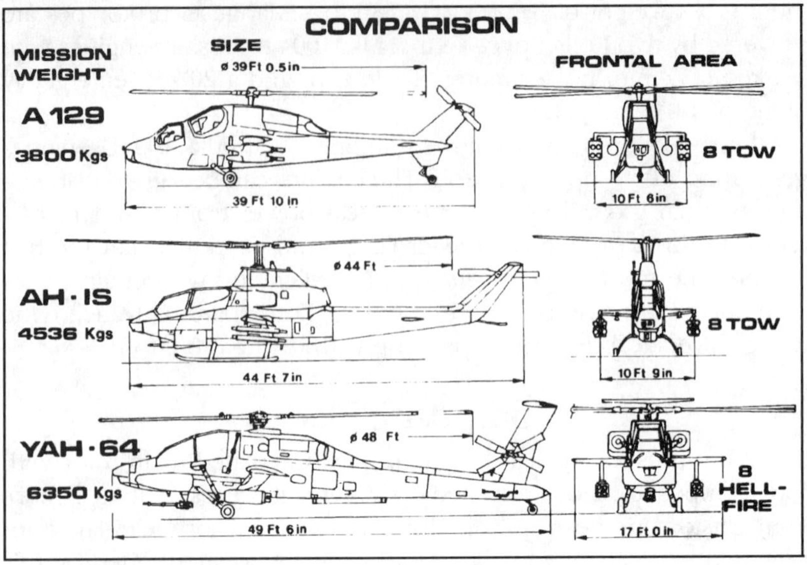

A129 is compared in size and weight with the Bell Helicopter Textron AH-1S Cobra and the McDonnell Douglas Helicopters YAH-64 Apache attack helicopter prototype in this Agusta chart. Production AH-64s are slightly heavier than the prototype.

gross weight. Tail rotor authority was more than enough to achieve high yaw rates through 360 deg. while hovering over a spot on the ground. The Italian army has specified a requirement for a 36-deg./sec. yaw rate to the left with no wind.

On production aircraft, the automatic flight control system (AFCS) will operate the rudder controls to provide coordinated turns, thus reducing pilot workload. The AFCS will normally be engaged at all times, and operation without the system will be considered a degraded control mode.

The army also has specified a 45-kt. capability in sideward flight. This can be achieved easily in sideward flight to the left, but only 42 kt. has been demonstrated to the right before maximum left rudder displacement is reached. Aircraft controls are responsive and precise in these maneuvers, much superior to the generally sloppy responsiveness of earlier generation helicopters.

Maximum power climb

I added power and accelerated to begin a maximum power climb at 65–70 kt. Our climb rate stabilized initially at 2,000 fpm, gradually decreasing to about 1,500 fpm as we passed through 8,000 ft. Aircraft weight at that point was approximately 3,800 kg. (8,360 lb.).

Leveling off at 10,000 ft., I established the aircraft in a 110-kt. cruise. Longobardi simulated an engine failure by retarding the power control lever for the No. 2 engine to ground idle. I brought the No. 1 engine to its maximum continuous power setting under those circumstances of 600C turbine inlet temperature, and this maintained our 10,000-ft. altitude at a speed of 75–80 kt. Increasing the power setting to the 2.5-min. limit of 635C TIT generated a slight climb of about 250 fpm.

The A129 has been cleared to an operational ceiling of 15,000 ft. thus far in its development, and some potential export customers are asking for even higher operational ceilings, according to Longobardi. However, this would require modifications to engine fuel systems and boost pumps. As Longobardi brought the No. 2 engine back on line, I entered a steep dive until airspeed increased to the A129's maximum allowable speed (V_{ne}) of 150 kt. Rotor-induced vibration level remained very low, although a mild 1-per-revolution beat did develop above 140 kt. Leveling off at treetop level, I established the aircraft at its normal cruise speed of 135 kt. using a power setting of 85–90% on the two engines.

Maneuvering load limits

At weights below 3,400 kg. (7,480 lb.), the A129 has maneuvering load limits of +3.5g and –0.5g. Above 3,600 kg., however, the aircraft is restricted to +2.8g maximum and +0.5g minimum. We were still above 3,700 kg. at that point, so I observed the tighter limits. Even these allowed a comfortable maneuvering margin to avoid obstacles at the low altitude we were flying.

Longobardi suggested a nearby golf course to examine the nap-of-the-earth flying qualities of the A129, and as I came across the uninhabited course, I reduced power and flared the aircraft for a rapid deceleration to a hover. Rotor speed remained constant throughout this maneuver, and the engines responded quickly to prevent any rotor rpm droop as I applied power at the end of the flare to bring the aircraft to a hover.

Keeping the aircraft below the treetops, I began threading my way around the golf course in simulated nap-of-the-earth flight at a speed of 10–20 kt. My caution at the controls of a new aircraft was obvious, and Longobardi took the controls to demonstrate that the A129's agility and maneuverability were such that it could be flown at this altitude within the relatively tight confines of the course at speeds of 40 kt. or higher.

One of the few unfavorable aspects of the A129 arose during these activities. This was the relatively high vibration level that developed at speeds between 20 and 40 kt. These have been measured at 0.81.8g, according to Longobardi. Although they do not affect safe control of the aircraft, the vibrations are uncomfortable and distracting enough that they could affect the outcome of missions. Fixes to the rotor system to correct this are under study. Longobardi said there are plans to modify the loads on the pitch control rods on the main rotor to determine if this will reduce the vibrations. If not, it may be necessary to change the mounts on the main transmission, he said.

At the conclusion of this demonstration, I took the controls again to perform a series of bob-up maneuvers, simulating bobbing up from behind a row of trees to fire at targets beyond them, and then ducking down behind the trees again. The responsiveness of the engines and flight controls permitted these maneuvers to be executed quickly and crisply. Attitude and heading remained fairly constant despite the major power changes, thus allowing target acquisition and simulated missile firing to be accomplished with minimum delay. Rotor rpm remained constant as well, and thus did not become a distraction during the maneuvers.

Automatic torque limiter

It was necessary to pay close attention to engine torque during these maneuvers to avoid exceeding torque limits, since this particular prototype does not have automatic torque limiters. On production aircraft, the automatic flight control system will limit torque automatically whenever the autopilot is engaged.

Before returning to the Agusta flight test facility, we diverted briefly to Milan's Malpensa International Airport adjacent to Cascina Costa where we performed several run-on landings with both engines operating and with one engine inoperative. The A129 has an operating limit of 40 kt. for ground operations, and at Longobardi's suggestion I aimed for a touchdown speed of 30 kt. in the run-on landings on the Malpensa runway. The aircraft could be landed smoothly with minimum power, whether from one engine or two. Purpose of the demonstration was to show that the aircraft could be landed safely and smoothly at high gross weights with minimum power available.

We landed at Cascina Costa shortly after this with a landing weight of 3,760 kg., having used approximately 160 kg. of fuel out of the 600 kg. with which we started. Total flight time was about 50 min.

The second Aviation Week flight in the A129 took place the following day, this time in the No. 5 prototype. This aircraft has a larger

A129 prototype No. 5 was flown on one of two Aviation Week & Space Technology *evaluation flights in the new attack helicopter. The aircraft is equipped with two large pods, one located under each stub wing, housing flight test instrumentation and data storage equipment.*

percentage of the integrated A129 avionics package than the No. 3 aircraft, and it was possible to examine some of the automatic flight control aspects of the helicopter.

Flight test pods

As in the first flight, our takeoff weight was 3,900 kg./8,580 lb.—the alternate maximum gross weight of the A129 for the Italian army. In addition to 600 kg./1,320 lb. of fuel, the aircraft also carried two large flight test instrumentation pods, one on each weapon pylon.

Startup was characterized by the same smooth rotor engagement and turn-up that we experienced on the first flight, and a relatively low level of rotor noise inside the cockpit. This remained unchanged throughout the flight.

This aircraft was equipped with a CRT multifunction display on the instrument panel in each cockpit, and it was possible to page through the pretakeoff checklists on the display.

An attempt is being made to keep checklists to a minimum, Longobardi said, and the pretakeoff checks will be limited to electrical and fuel system items. On production aircraft, mission-related systems will be checked automatically by built-in test equipment. The

BIT equipment will be connected to a data transfer unit into which the crew will insert a cartridge containing specific mission details.

With no mission data to enter or check for this particular flight, we finished the checklist quickly and lifted off for the second flight. As I held the helicopter in a 20-ft. hover near the flight line, Longobardi turned on the stability and control augmentation system (SCAS) to demonstrate the workload reduction made possible by the system.

I took my hands off the controls. The system maintained our attitude, heading and altitude, but the aircraft began a slow drift to the right. As with other such rate stabilization systems, the A129 SCAS does not always respond to those changes that occur at a very slow rate. Such was the case here.

I engaged the Doppler hover mode on the control display unit in the front cockpit. The Doppler system brought the aircraft to a stable hover, eliminating the right drift that had developed. In normal operations, the hover mode will be engaged when the aircraft's speed drops below 10 kt., Longobardi said. As speed drops below 2 kt., the system memorizes the aircraft's position at that point and begins working to return and hold the helicopter over that spot.

I disengaged the hover mode and pushed the "ATT" button on my control display unit to select the attitude hold mode of the automatic flight control system. The system held the aircraft reasonably

Another A129 prototype shows the agility of the helicopter in a near-inverted maneuver during a flight demonstration.

steady in this mode, although it was subject to external forces such as wind gusts that occasionally tipped the rotor slightly, necessitating occasional small control inputs on my part to maintain the hover.

With attitude hold engaged, it was possible to make small power changes without the torque variations causing significant changes in heading. Specifications for the A129 SCAS include a requirement to maintain heading when torque is varied by up to 10% without the heading hold mode engaged. In this case, I was able to increase and decrease torque by up to 20% without causing any significant heading changes. After engaging both the hover mode and the heading hold, I could make unlimited power changes with my feet off the rudder pedals while the automatic flight controls kept the heading constant.

Using a beep trim switch on the collective pitch control lever, I was able to increase or decrease the helicopter's altitude at a comfortably slow rate. Pushing the same switch to the left or right caused the aircraft's heading to change in small increments in the direction the switch was pushed. The A129's precision aiming capabilities were evident in this brief demonstration.

Hovering at maximum gross weight at the Cascina Costa facility's 700 ft. elevation with the temperature still well below standard required about 82–83% torque on the two Gem engines, leaving a substantial power margin. I pushed the nose down slightly and applied more power to accelerate to forward flight. The A129 has an acceleration requirement from the Italian army of 0–40 kt. in 5 sec., and the aircraft easily meets this requirement.

As our ground speed surpassed 10 kt., the hover mode disengaged automatically. I selected the "flight plan" mode on my control display unit on the left console, and the multifunction CRT showed a horizontal situation display with our flight path along a route that was programed in by Longobardi prior to takeoff. The scale of the HSI map display could be selected in increments from 5 km. to 160 km. to show smaller or larger segments of the route as desired.

Waypoint distance

I let our airspeed stabilize at 110 kt. as we cruised above the Italian countryside at an altitude of 2,200 ft. The electronic horizontal situation display showed a distance of 22 km. to our first waypoint, the Saronno VOR, and it also indicated a time of flight of 8 min. at that speed. Flight director bars on the attitude director indicator provided steering cues to the waypoint. Production aircraft will be equipped with an automatic pilot capable of being coupled to the Doppler navigation system for a hands-off flight capability in normal cruise conditions. The autopilot was not yet fully operational on this aircraft.

The Mangusta's two-blade, wide-chord tail rotor operates at low tip speeds to reduce noise.

I rolled the aircraft into a left turn, then released the stick. The attitude hold mode of the flight control system held the aircraft in the turn with the attitude I had selected before releasing the stick. This is the equivalent of a control wheel steering (CWS) mode on a transport aircraft. Likewise, when I pulled back on the stick slightly and then released it, the aircraft maintained the new nose attitude I had selected. Several more maneuvers of this type demonstrated the stability that could be expected from the A129 under instrument meteorological conditions. This will be an important factor in an aircraft that is required to be operable at low altitudes at night and in adverse weather.

We returned to Cascina Costa along the same route, and I increased the speed to 135 kt. for another demonstration of the A129's low vibration levels. After landing at the flight test facility, I taxied to our parking space. The aircraft handles quite easily on the ground with its forward-mounted nonretractable main landing gear and swiveling tail wheel.

Outside the aircraft, its low rotor vibration levels translate into low noise levels as well. In a flight demonstration presented the same day by Agusta test pilot Fabio Frisi for a group of Italian army and Defense Ministry officials, the low noise of the A129 allowed Frisi to fly it from one end of the airfield to the other behind a treeline a few hundred feet away without detection by those standing on the flight line.

Full capabilities of the A129's integrated multiplex system could be seen in a laboratory "hot mockup" of the system inside a hangar here. The systems and software integration device, complete with an A129 cockpit, has provided the capability to accomplish some limited closed-loop testing of the A129's systems with a pilot or technician operating the aircraft's controls and systems.

Central processing units

The integrated A129 system, designed by Agusta and the Harris Corp., goes beyond MIL-STD-1553 data bus architecture, according to program officials, providing integrated controls and displays plus central computing redundancy. Five central processing units are used to coordinate and monitor the operations of all communications and navigation equipment, fuel/electrical/hydraulic systems, flight controls, engines, transmission, electro-optical sensors and weapon systems.

Fully integrated in the aircraft, the system performs a variety of functions, including maintenance troubleshooting and data recording, navigation, weapons delivery guidance, and engine and performance monitoring. Selecting the "attack" mode during a mission will provide the pilot with pitch and roll steering cues on his helmet-mounted sight for the accurate firing of rockets. It is anticipated that the pilot will normally fire the rockets, while the copilot/gunner will fire the TOW missiles or other such guided antitank weapons. Program officials said the system will permit the crew to fire rockets on a ballistic trajectory without actually seeing the target if accurate target data are available from another source.

Provisions are also included in the A129 for a chin turret gun, with several options available to potential customers. The Italian army has specified a .50-caliber machine gun, but other potential cus-

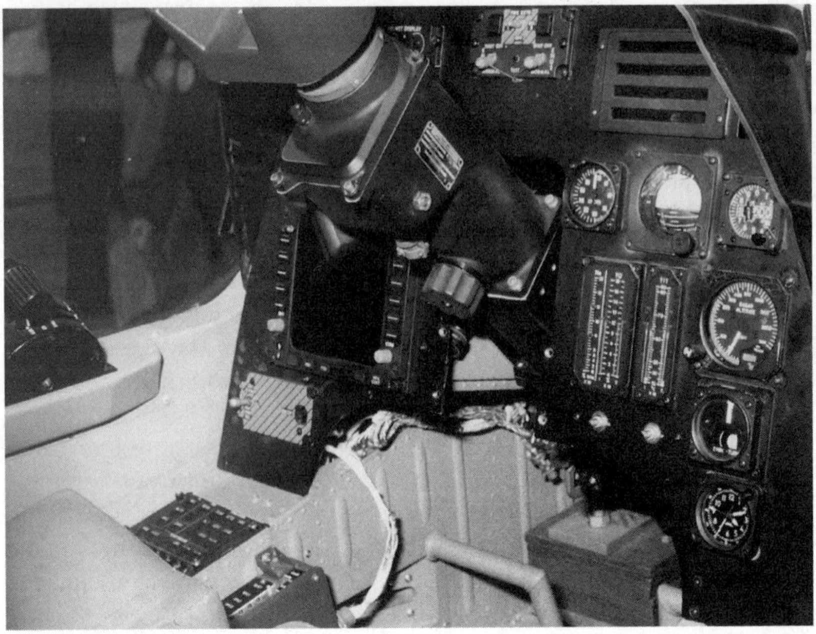

Front cockpit of the A129 helicopter has a single multifunction CRT display on the left side of the instrument panel. Front cockpit is also equipped with a TOW missile sight for the copilot/gunner.

tomers are expected to be satisfied with a 12.7-mm. cannon. Program officials believe the accuracy and fast response of the 12.7-mm. gun will make it equally as effective for air-to-air and air-to-ground applications. They also believe the emerging requirement for an air-to-air capability in attack helicopters will require a gun capable of firing at a rate of about 1,500 rounds per minute.

Rather than having to calculate performance margins in the A129 with varying temperatures, altitudes, weights and ordnance configurations, the pilot will be able to call up a power performance index on his multifunction display that will provide an indication of engine margins under the prevailing situation. A performance monitor mode can be selected by the pilot to provide information such as hover capability, cruise capability, power available, time on target, and one-engine-out performance. For night operations, the A129 will be equipped with a Honeywell forward-looking infrared system (mini-Flir) mounted in a turret in the nose of the aircraft. The copilot/gunner in the front cockpit will use his M-65 TOW sight to view the Flir imagery while the pilot in the rear cockpit will have the Flir imagery displayed on his Honeywell IHADSS helmet-mounted sight.

Rear cockpit of the A129 helicopter also has a single multifunction CRT display on the left side of the instrument panel.

Target designation system

Program officials earlier had considered using the Martin Marietta target acquisition designation system/pilots night vision system (TADS/PNVS) used on U.S. Army/McDonnell Douglas Helicopters AH-64 Apache attack helicopters, but chose the Honeywell Flir system instead because of its lower weight and because they believed the TADS/PNVS was optimized for the AH-64. Provisions have been incorporated in the A129, however, for the installation of a mast-mounted sight above the A129 rotor for those customers who prefer a system of that type.

Agusta officials have been examining other applications for the A129 and its derivatives, including antiship, scout/close air support, air defense, light battlefield support and search and rescue versions. The latter two would require new fuselage designs but would use the same drive train.

Since the beginning of this year Agusta engineers, along with those of several other European aircraft manufacturers, have been studying a growth version of the A129 called the Tonal (AW&ST Jan.

A129 prototype fires a Hughes Aircraft Co. TOW antitank missile during flight testing of the helicopter. Initial production A129s will be configured to carry eight TOWs or 52 2.75-in. rockets, but provisions are included for weapons system growth to accommodate a new generation of antiarmor weapons, air-to-air missiles and guns.

Light weight, high power and an advanced flight control system give the A129 exceptional maneuverability and responsiveness at high speeds and low altitudes.

19, p. 78). An intermediate report is due early next year, and a final report is scheduled to be submitted in early 1989.

Fire-and-forget weapons

Program managers at Agusta believe about 70% of the aircraft will remain unchanged, and the primary difference will be in the integration of new weapon systems. The aircraft will be equipped with fire-and-forget antitank weapons, allowing it to operate at higher speeds and to maneuver immediately after launching its weapons.

One of the other European manufacturers involved in the studies is Britain's Westland, and Agusta officials believe the Tonal will employ the Westland-developed BERP rotor blade design that permitted a Westland Lynx equipped with the blades to set a new world speed record for helicopters last year. All of the improvements considered for the Tonal will be tested first on the A129, according to program officials, and will be developed in retrofit form so they can be applied to the A129 as well.

Chapter 2

Commercial

MD Explorer readied for certification

David L. Green/Mesa, Ariz.
August 29, 1994

The civil market awaits the first three production aircraft, which are expected to be FAA-certified and ready for delivery in December.

The McDonnell Douglas Helicopter Systems Explorer offers civil operators a seven-passenger, twin-engine helicopter developed for single-pilot operations. It is positioned to compete for the emergency medical services business.

The 6,000-lb. Notar (No Tail Rotor) antitorque and directional control system-equipped Explorer, capable of lifting 3,000 lb. from a sling lift or carrying a 1,600-lb. payload, is priced at $3 million for the basic aircraft. The helicopter also offers a large cabin, quiet performance and easy maintenance.

The Explorer is expected to be the first helicopter to receive FAA production approval of a hingeless, bearingless rotor. The Explorer also marks the first installation of a Pratt & Whitney Canada PW-206A turboshaft engine. Certification and delivery of the first three production aircraft are scheduled for December. The fourth production aircraft will be an EMS demonstrator, and is expected to be on display at the National Business Aircraft Assn. convention in New Orleans (AW&ST Oct. 4, 1993, p. 47).

The *Aviation Week & Space Technology* evaluation flight in the production-configured MD Explorer Unit No. 10 was flown with company pilot Chan Morse at the company's main plant here. We operated the basic VFR aircraft at a takeoff weight of 5,700 lb. under hot day, no wind conditions.

The Explorer's rotor does not have mechanical hinges or lead-lag dampers, but does have a relatively large, effective offset hinge of 5.5%. Low maintenance cost, excellent controllability and a robust flight envelope are a product of the rotor design.

Rotor loads are carried directly from the rotor head to two major fuselage frames, one forward and one aft of the main cabin. This arrangement is patterned after the MD 500 and AH-64, and allows the transmission to be removed without disturbing the rotor system or rerigging the upper control system.

The rectangular 172-cu.-ft. cabin is located between the two main frames with a 146-gal., crash-resistant elastomeric fuel cell below the

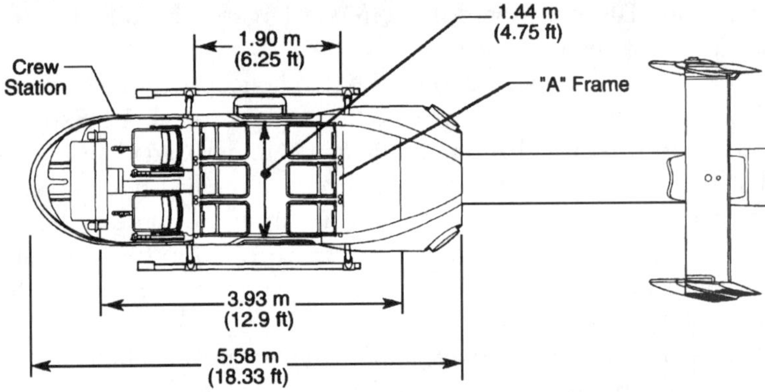

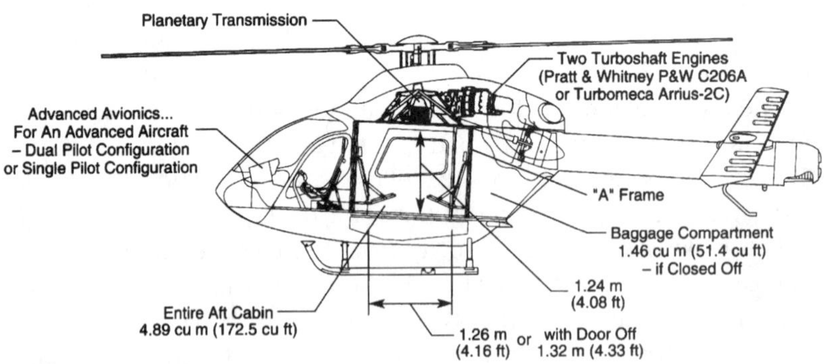

The first production MD Explorer is scheduled to fly soon and will be delivered to Petroleum Helicopters, Inc., in December. McDonnell Douglas Helicopter Systems has 200 firm orders.

deck and a rearward sliding door on either side of the cabin. A 48-in.-wide pallet can be loaded on the flat cabin deck and 14-ft.-long tubing fits inside the cabin when the copilot's seat and controls are removed.

Main frames are rigidly attached to the cross tubes of the landing gear. The aft cross tube connects to the skids via dampers that protect against ground resonance that might otherwise be induced by the rotor.

Swinging into the seat through the right door is facilitated by the shape of the cyclic, which is bowed forward so that one can swing a leg under its grip. This curvature minimizes the possibility of the pilot's leg interfering with the displacement of the stick during aggressive maneuvers. In addition, the control pivots about a point below the seat itself, causing the cyclic grip to move downward slightly with forward tilt, accommodating the desire of many pilots to rest their hands on their legs as the stick position changes with center of gravity or speed. The Explorer's seat design accommodates the ergonomics of both tall and short pilots, ensuring each has a good view of the instrument panel and over the nose.

The MD Explorer's cockpit is well arranged and uncluttered, with excellent forward visibility. A five-color liquid crystal integrated instrument display system presents fuel, engine, rotor and systems information including on-board health monitoring. Full authority digital engine controls (FADEC) and manual throttles incorporate important innovations for training and single crew emergencies, while hydraulic failures and other system malfunctions require little, if any, pilot intervention.

I centered the stick, checked the collective down and prepared to start the engines. The aircraft can be started and readied for flight in less than a minute. The engine start sequence involves checking the rotor brake "off," turning on the generators and fuel boost pumps. The next step is to lift and rotate one of the two power management controls to "idle." Sequencing of the starter-generator, ignition and fuel scheduling is automatic. The rotor started to turn with engine ignition then stabilized at 65%.

The sequence was repeated for the second engine, and the power management controls were both placed in the "fly" position, producing a rotor speed of 101% in well under a minute. Hydraulics pressurize when the rotor turns; there are no "off-on" checks to complete prior to flight. The integrated instrument display system (IIDS) monitored the start-up process, and if there had been anything wrong, the on-board health monitoring system would have alerted us. Morse called for clearance and we were ready to lift. We lifted into a hover, and I air-taxied to a nearby taxiway to conduct low-speed controllability evaluations and was pleased with the docile flying qualities in the slow speed envelope.

The flight then proceeded to a grassy area nearby and I flew patterns for steep approaches and departure from a hover. The hover work in the Explorer included landings and single-engine emergencies. While airborne, the Explorer controls are responsive and do not present the pilot with any surprises.

Cyclic control forces were light and well harmonized. Breakout forces at trim were light and well-tailored, making it easy to move the control smoothly from one side to the other.

Patterned after the MD 500, cyclic stick centering involves a "spring force feel" system, which is controlled by a four-way trim switch, but there is no momentary trim release button.

The collective is held in place by 20 lb. of friction until turned off or momentarily released by a switch mounted on the underside of the collective. I found the collective to be balanced with friction deselected.

Left lateral control was needed to roll left as the aircraft lifted off left side down. This lateral control deflection was substantial, but not uncomfortable, and left pedal was required for takeoff. The initial response of the aircraft to small pedal inputs is much like that I have experienced in tandem rotor helicopters. The response is softer than the sharp edge response of a conventional tail rotor, and should provide the passengers a smoother ride in turbulence.

Maneuvers included high-speed rearward, sideward, and turning translational flight to simulate operating in wind conditions in excess of 30 kt. I never approached a control limit or encountered an adverse interaction between the main rotor and the vertical or horizontal tail.

Lateral and rearward flight produced no sudden pitch, roll or yaw departures, although cyclic control displacements were substantial during some accelerations and translating, turning maneuvers. Landings from a hover were level, fore and aft, with left side down. There was no adverse ground effect nearing touchdown.

Next, I explored the training function of the power management system. Morse rotated the No. 1 engine control to TRAINING and the engine decelerated as though it had failed, while the No. 2 engine took the load and maintained rotor speed at 100%. The FADEC had actually reset the power turbine speed to 92% on the No. 1 engine.

If an engine malfunctions, the deselected engine will automatically come to full power and revert to the original 100% Nr (main rotor rpm) governing schedule, an excellent form of imbedded training for practicing engine failure flight techniques during slow speed operations.

The Explorer cockpit provides outstanding visibility and is uncluttered and well-arranged for single-pilot operations.

IIDS DISPLAY FORMAT - LAMP TEST MODE

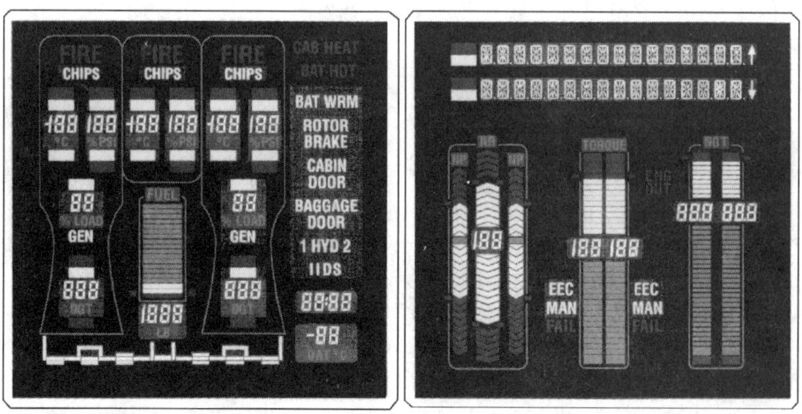

The Integrated Instrument Display System provides power management instrumentation and health monitoring data.

Strong customer inputs prompted McDonnell Douglas Helicopter to design an engine control that permits a pilot to gain manual control instantly of both engines in the unlikely event of a directional control problem. Twist grip manual throttles on the Explorer's collectives respond to this requirement. Rotating a twist grip immediately places the engine in a manual mode and links the twist grip to a DC stepper motor, which controls the engine main metering valve. The remaining engine provides rotor rpm governing until it either reaches full power or zero power.

I explored the transition to manual control and concluded that the task is easy and can be accomplished without letting go of the collective. While in a hover, indicating 65% torque of both engines, I rotated the No. 1 throttle ever so slightly. Manual control was achieved immediately, as I smoothly reduced torque to 30% with 20- or 30-deg. of rotation. The PW-206A engine response was quick but well damped. The relationship between the twist grip control and the engine was ideal. Depressing a reset switch positioned just forward of the No. 1 engine twist grip causes the engine to revert to normal control.

An unhurried departure from the hover into forward flight required almost no pedal adjustments between the pushover and gaining 50 kt. or so in a climb. A more aggressive departure seemed to require more pedal trimming as speed built up and the vertical fins became effective. Depending upon the acceleration technique, there sometimes was a need for forward cyclic to keep the nose down, and other times the control moved very little during the acceleration.

The improved Notar directional control system makes the Explorer easier to control in hover than earlier versions. The absence of a turbulence-sensitive tail rotor and the "soft" response of the directional control thruster to pedal inputs enhance the ride in the main cabin. A yaw stability augmentation system (SAS) and a collective-to-yaw coupling system work through the vertical fins to reduce pilot workload and prevent Dutch Roll tendency with a two-second period, which Morse said the aircraft exhibits with the SAS turned off.

I climbed the Explorer to the north of the plant to 3,000 ft. to evaluate handling during climbs, descents and turning maneuvers. The damping had been observed to lessen as speed decreased and power was increased during a climb. Coordinated turns in climbing flight were easy to establish and maintain. There was no Dutch Roll tendency with the SAS on.

The SAS works through the Vertical Stabilizer Control System (VSCS), a dual channel system that combines the input from the SAS with a second signal from the pilot's collective. The combined signal controls the incidence of the vertical tails through electro-mechanical actuators located inside the horizontal stabilizer. Each fin has its own VSCS channel.

The signal from the collective control deflects the vertical tail surfaces to compensate for variations in torque as the collective is moved up and down. This provides a collective-to-yaw coupling intended to decrease the pilot's workload during constant speed power changes, and provides the antitorque moment needed for steady-heading balanced flight at cruise so that the pedals can be neutralized.

The rear hatch provides access to the baggage compartment and the 172-cu.-ft. main cabin area. This access allows loading of 14-ft.-long tubing when the copilot's seat and controls are removed from the helicopter.

The grip and curvature of the cyclic control are unique. The curvature minimizes the possibility of the pilot's leg interfering with the displacement of the control. It pivots about a point below the pilot's seat, allowing the grip to move slightly downward as the stick tilts forward.

This pedal position minimizes the power demanded by the Notar fan, with the effect of maximizing range. The system works well for fixed speed power changes, but you need to keep flying the pedals during constant speed power changes.

I then pushed over into a shallow dive slowly to attain maximum speed. The Explorer picked up a substantial right sideslip as the airspeed passed through 120 kt. Passing 135 kt., I put the aircraft back into balanced flight with about 3 in. of right pedal.

According to McDonnell Douglas, the 6,000-lb. helicopter has a useful load of 2,565 lb. (1,163 kg.) and a maximum cruise speed of 145 kt. (268 km./hr.) at sea level. With a useful fuel capacity of 159 gal. (600 liters), the range is 315 naut. mi. (583 km.) at 5,000 ft. The maximum endurance for the MD 900 is 3.5 hr. The Explorer will hover out of ground effect at 11,300 ft. (3,444 meters) on a standard day.

The Explorer design, designated the MD 900, was aided by a company worldwide survey involving 200 operators. McDonnell Douglas established a Blue Team supported by 15 operators that helped it to refine the Explorer's design.

The sensitivity to direct operating cost prompted McDonnell Douglas to retain proven design concepts and introduce new technology only where it paid for itself by reducing weight, increasing reliability or increasing payload, Dean Bergman, McDonnell Douglas Helicopter senior vice president, said.

The flight development program began with the Explorer's first flight in December, 1992. The earlier Explorer test flights were flown with a PW-206A engine with a 640 eshp. rating. Four MD 900s have accumulated 700 flight hours and have been cleared of all the airworthiness issues for a normal gross weight of 6,000 lb., or 6,900 lb. with a sling load. The performance sought by McDonnell Douglas dictated a higher 720 eshp. rating. The Explorer flown had the higher rated engines. The transmission was also uprated and now has about 1,400 flight and ground test hours.

Hydraulic boost for the main rotor was an important issue in flight test. The 6,000-lb. weight was just high enough to bring into question the advisability of normal operations without boost. The original design included provisions for hydraulic boost in the instrument flight rules configuration, but company officials had hoped to avoid the weight and cost of a hydraulic system in the basic helicopter.

Company pilots found that unboosted flight was impressive, but time for refinement was short and a dual, 500-psi., irreversible single-pilot system was installed. The pumps, reservoirs and actuators are very compactly arranged forward on the transmission deck. Failure of one subsystem is transparent to the pilot and the faulty system is shut down without pilot intervention.

The hydraulic system also powers a single actuator, which controls the Notar fan and rotation of the thruster. Engines are mounted aft of a single transmission, one on either side of the Notar fan and

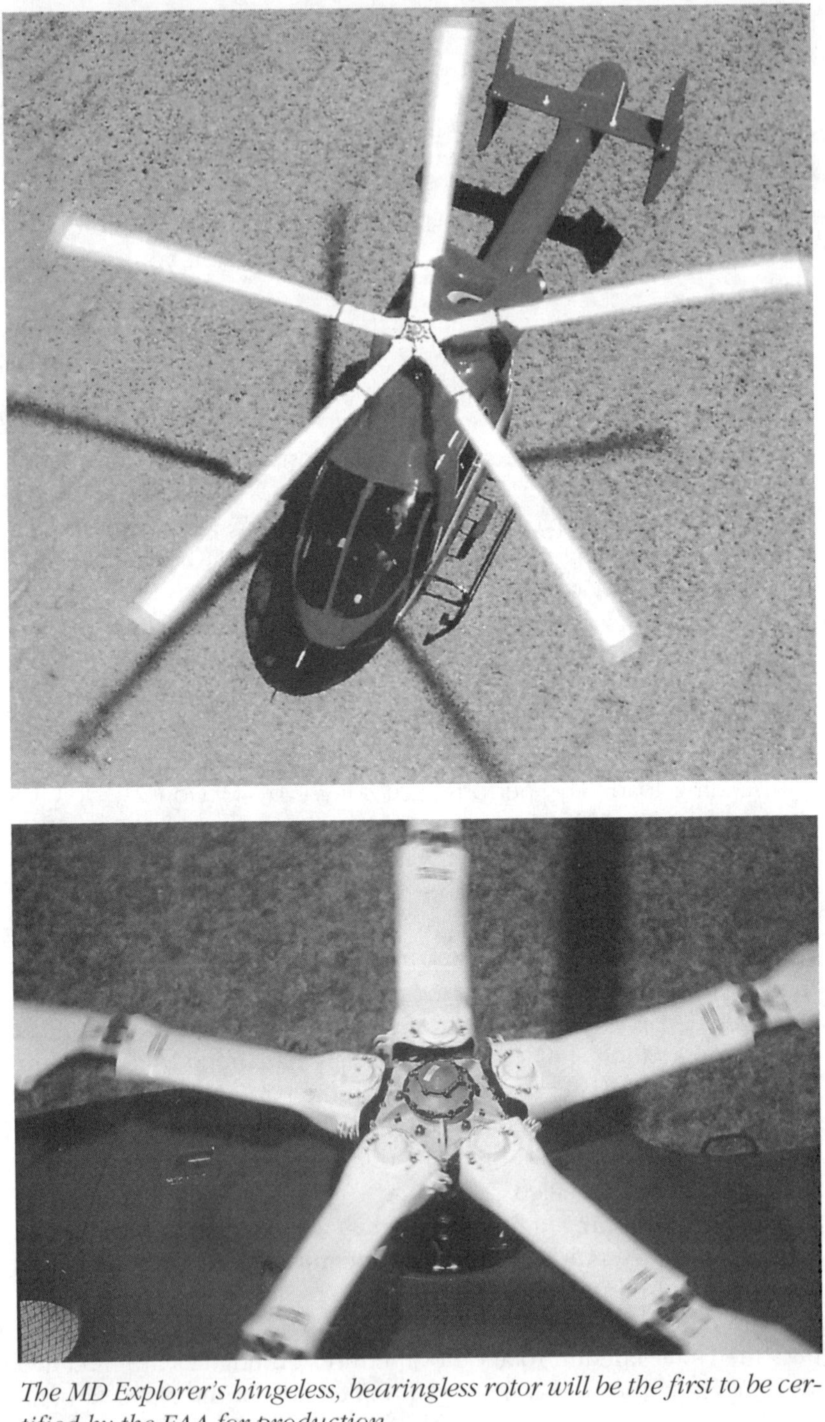

The MD Explorer's hingeless, bearingless rotor will be the first to be certified by the FAA for production.

inlet. Two horizontal Coanda slots are found on the lower right side of the cylindrical tailboom, which incorporates an "H" tail with controllable vertical fins.

Early Notar thruster designs used turning vanes on the left, antitorque side of the thruster, but on the pro-torque side of the thruster there was a single open passage. Under certain conditions pilots reached control limits, Evan Sampatacos, the company's designer of the Explorer, said.

This characteristic was eliminated during the MD 520 program by incorporating turning vanes on both sides of the thruster. In addition, the MD 900 incorporates a nonlinear cam in the directional control system to enhance yaw moment response on either side of neutral.

The Notar is improved over earlier versions and incorporates turning vanes on the right, antitorque side and a special cam in the tail-rotor control system to enhance the directional control scheduling. The resultant control characteristics are quick but very smooth, eliminating the sharpness of the typical tail rotor to pedal inputs and gusty wind conditions. The smooth response provided by Notar directional control system also improves the ride in the main cabin. A stability augmentation system works through the vertical tails and eliminates a Dutch Roll tendency inherent in the basic aircraft.

With the improved thruster, the helicopter flew fine except for a Dutch Roll characteristic that was eliminated by yaw rate damper.

McDonnell Douglas Helicopter Systems currently has 200 orders. The first production MD 900 is scheduled to fly in the third quarter of this year, and will be delivered to Petroleum Helicopters, Inc., in December. The company's order book indicates that the strongest current interest in the Explorer is from the EMS sector.

Piston-powered R44 offers low-cost flight

David A. Brown/Ft. Worth
March 29, 1993

Robinson expects a two-per-day production rate for its four-seat R44—the quiet, more spacious corporate helicopter of more advanced design than the R22.

The Robinson R44 four-place corporate helicopter combines state-of-the-art electronics with the latest versions of airframe and drive system designs from earlier Robinson rotorcraft to provide turbine-engine performance at piston-engine cost. The helicopter also offers near turbine vibration levels and high-speed cruise.

The first production model of the R44, N445RH, was flown here recently by this *Aviation Week & Space Technology* pilot during its stopover on a transcontinental flight from Miami Beach, to the company's home base at Torrance, Calif. The helicopter, flown by Robinson's chief pilot, Douglas Tompkins, was returning home after being displayed at the Helicopter Assn. International Convention. He was accompanied by Patrick Cox, development engineer on the R44 project.

The R44 design, based on the company's experience with the earlier two-seat Robinson R22, has been certificated under Federal Aviation Regulations Part 27, including amendments 1-2A, and FAR Part 36,

First production model of the Robinson R44 four-place corporate helicopter flies low past the control tower at Ft. Worth Alliance Airport.

including amendments 1-20 using Appendix J procedures. It currently is certificated for day and night visual flight rules operations, and there are at present no plans to certificate it for instrument operations. The Federal Aviation Administration awarded the R44 its type certificate on Feb. 15, and the company says it had in excess of 135 orders for the helicopter as of the close of the HAI meeting on Feb. 27.

A new factory is under construction at Torrance to keep the production rate up to the level that the company anticipates will be necessary, but it will not be completed until late this year. Until the factory is in full operation, the production rate of the R44 will be limited to approximately 1.5 per week, and on this basis, production is sold out through the end of 1994. Most of the first year's production will go to Robinson distributors for use as sales demonstrators, although some operators will also receive delivery of R44s.

Eventually the company hopes to boost production to two R44s per day, along with an equal number of the R22, if demand warrants. Although similar in appearance and with some of the same basic design features as the R22, the four-seat R44 is a considerably more advanced design.

The helicopter is powered by a Textron Lycoming O-540-F1B5 engine derated from its normal 260-hp. output to a five-minute take-off rating of 225 hp., and a maximum continuous rating of 205 hp. This is done via a simple cross-reference placard in the cabin, which

Unusual T-Bar cyclic control system permits lighter and simpler control layout. Note compact, centrally mounted instrument console to permit greater external visibility.

Tail of the R44 has an offset vertical fin to unload the tail rotor in forward flight.

Large striped guard protects tail rotor when landing in rough terrain.

indicates the maximum manifold pressure that can be utilized at any given combination of outside air temperature and pressure altitude.

The engine is normally aspirated, and the derating is designed both to provide an ample reserve of power under all feasible operating conditions and to save on engine wear, according to Cox. The engine drives the main and tail rotors using a modernized version of the belt-drive system in the R22. An electrically powered, spring-loaded alignment system assures that the drive shafts are properly aligned when the engine is operating. The system also permits the drive system to be retrimmed in flight when, for example, the belts in the main drive system expand because of heat.

Only flex couplings are used in the power-train system. There are no universal joints or gear couplings. The flex couplings do not require maintenance during the 2,000-hr. time-between-overhaul (TBO) for which the R44 is approved. All life-limited components have service lives of at least 2,000-hr.

The two-bladed rotor system uses a three-hinge design that does away with conventional lag hinges, hydraulic struts and dampers. The rotor operates in a range of 392–404 rpm (99–102% of nominal) with power on, and in a range of 356–428 rpm (90–108%) in autorotation.

The rotor and engine rpm are controlled by an electronic rpm governor that holds both speeds constant even in such adverse conditions as high wind gusts, which normally cause rotor speed to fluctuate noticeably. The governor engages when the engine and rotor reach 80% nominal speed and is designed to automatically advance both to 100% and hold them there to within plus or minus 1%.

Rapid opening access doors on the R44 permit quick inspection of the helicopter's drive system and key engine components.

This allows the pilot to fly the helicopter as if it were a turbine-powered aircraft and eliminates the continuous throttle twisting that was necessary on older piston-powered rotary-wing aircraft. It also eliminates the need to have an ear attuned to engine sounds to help detect an underspeed or overspeed condition. This proved to be an advantage because the engine muffler and cabin acoustical buffering tend to reduce the engine sounds considerably in the cabin. The tail rotor also is designed to have a low tip speed to further cut noise.

The rotor engagement is controlled by an electric clutch, which requires only that the pilot flip the guarded switch to engage the rotor after the engine is started. There also is an automatic start sequence cutout to prevent the rotor from being engaged automatically at engine start if the switch is inadvertently left on, and there is provision for disengaging the clutch in flight to permit an in-flight start of the engine.

I flew the R44 from Ft. Worth Alliance Airport on a blustery day, with the ambient temperature recorded at 63F and the wind from the west-northwest at 25 mph with gusts exceeding 30 mph. The helicopter had 67 hr. total time at the time of the flight.

Tompkins showed me the access panels that quickly opened on each side of the engine compartment to give quick access to the oil quantity gauge and to inspect the belt drive system. For a complete view of the engine, it is necessary to get under the engine compart-

Clean rotorhead design on the R44 has eliminated many conventional components, including lag hinges, dampers and hydraulic struts.

ment and look up, but this is not necessary except for the first flight of the day or in unusual circumstances.

Two most noticeable features of the R44 cabin are the large viewing area—both from the front and rear seats—and the unusual center-mounted cyclic control stick and its projecting horn-shaped, TBar handle. This permits either pilot to operate the helicopter from a single collective control and is designed to facilitate ease of entry into and exit from the front seats.

The single collective concept has a more basic function in that it permits the helicopter to have a lighter and simpler control system. The control system uses only bell-cranks and push-pull tubes, with all control cables and pulleys eliminated.

Tompkins, from the left seat, ran through the simple and standard prestart sequence and started the engine, using the ignition key, just like the driver of a sports car. In fact, the R44 has a lot of features that could have been lifted directly from a medium-priced sports car. If it is not a Ferrari, it is at least an MGB-GT.

We let the engine idle until the cylinder-head temperature indicator was in the green, and then Tompkins engaged the electric clutch switch. The rotor accelerated at a moderate speed, and when the engine and rotor needles married at about 70%, Tompkins engaged the

Two cabin doors on each side of the fuselage permit easy access and exit. Note baggage stowed on rear seat. Helicopter also has underseat baggage hold.

rpm governor and held the throttle until both engine and rotor had reached 80% speed. At this point, the governor took over and advanced both engine and rotor to 100%.

We lifted into a hover with the engine producing about 19 in. on the manifold pressure gauge, and Tompkins demonstrated the R44's hovering capability, then let me fly the aircraft. I found the R44 to be lighter on the controls than I had anticipated, but placement of the cyclic control, which hung down from the TBar, rather than protruding up from the floor, was most disconcerting. I tended to fight the system, rather than letting it stabilize, and Tompkins noted that this was frequently the case with pilots unfamiliar with the concept. In two or three flight hours, most pilots become accustomed to it and find it no more unusual than the governor-controlled collective, he said.

There is a small thumb trim on each cyclic handle, and Tompkins demonstrated that the R44 could be trimmed to hover hands off, even in the high wind conditions we were experiencing.

We took off from Alliance, climbing out at the helicopter's best climb speed of 60 kt., which gave a rate of climb of just under 1,000 fpm. Vibration levels were considerably less than I had anticipated, although some rotor vibration tended to feed back through the cyclic. We made a 280-deg. climbing turn to the right to clear traffic, then departed the airport to a clear area about 7–10 mi. to the northwest, where I flew a figure eight to get accustomed to the R44's handling characteristics.

Heavier gusts tended to bounce the helicopter—which has a gross weight of 2,400 lb.—but the governor held rotor rpm virtually steady at 101–102% under all flight conditions and all angles into or out of the wind. Our takeoff weight was just over 2,300 lb., including nearly full fuel. The tank holds 48.9 gal. of usable fuel, which is fed to the engine by a gravity flow system that does not use pumps or fuel injectors. Empty weight of this particular R44 was 1,430 lb.

The R44 accelerated rapidly to 120 kt. indicated airspeed (IAS), which is close to the maximum speed at which the helicopter can operate, with the engine delivering 22 in. of manifold pressure. Never-exceed speed at weights over 2,200 lb. gross weight is 117 kt. calibrated airspeed (CAS). For weights under 2,200 lb., this increases to 127 CAS.

Tompkins noted that on the R44's flight from Torrance to Miami Beach, the average ground speed was 110 kt. with most of the flight conducted into a light 8–10 kt. headwind. Flying time from Torrance to Miami Beach was approximately 21 hr. 30 min. at 115–118 mph

Head-on view of the R44 shows clean lines and low-drag fuselage, which contributes to the helicopter's speed capability.

IAS. The engine, which operated about 22 hr. 30 min. during the trip, burned less than two quarts of oil during the trip, he said.

We shot an autorotational descent from approximately 1,400 ft. altitude to an empty field, stabilizing the airspeed at 70 kt. after entering autorotation. The R44 held a nearly level attitude during the descent and did not vary its attitude more than 2–3 deg. as power was added to arrest the helicopter's descent at an altitude of approximately 100 ft.

After we returned to Alliance, Tompkins again demonstrated the R44's autorotational capabilities by shooting an autorotational landing to the main runway. He noted that the R44 had a tendency to float, which the pilot had to be aware of and compensate for to avoid overshooting the intended touchdown point. He killed forward speed with a moderate flare and lowered the helicopter to the runway from about a 10-ft. hover height.

Two additional autorotations from hover further demonstrated the R44's capability to take advantage of ground effect in cushioning an unpowered touchdown. The R44 is equipped with a rotorbrake, which is applied by pulling an overhead T-handle located between the pilots' seats.

Visibility from the R44 in flight is exceptional, even downward and rearward. There is no bulkhead or other viewing impediment between the front and rear seat pairs. The pilot's possible tendency to look at the scenery is compensated for by an array of warning systems on the centrally mounted instrument console, which include

Robinson R44 cabin frame structure incorporates energy-absorbing landing gear to cushion vertical impact. Cabin has fore and aft-stressed seat-belt system.

main gearbox temperature and chip warning lights, a tail gearbox chip light, and lights to warn of low fuel, low oil pressure, low alternator voltage and rotor brake and engaged starter. A light and horn warn of low rotor rpm.

The R44 also overcomes some of the passenger comfort problems inherent in small aircraft. It has a pressure ventilation system that keeps the air fresh without any sense of air blowing on the crew or passengers. Pilots of more than average size do not feel particularly cramped in the cabin, although leg room in the rear seats is limited. Baggage can be stowed under the rear seats, with a limit of 50 lb. in each compartment. With less than four persons traveling, baggage also can be strapped down on the rear seat.

ROBINSON R44 HELICOPTER SPECIFICATIONS

Powerplant	Lycoming O-540-F1B5 derated to 205 hp.
Gross weight	2,400 lb.
Empty weight	1,400 lb.
Fuel, standard	31.6 gal./190 lb.
auxiliary	18.3 gal./110 lb.
Payload with standard fuel	810 lb.

Rotor diameter	33 ft.
Tail rotor diameter	4 ft. 1 in.
Max. height to top of rotor	10 ft. 9 in.
Max. airframe width, excluding rotor	8 ft. 2 in.
Cabin width, external	4 ft. 2.5 in.
Never-exceed speed, 2,200 lb. and up	117 kt. CAS
under 2,200 lb.	127 kt. CAS
power off	97 kt. CAS
Max. operating altitude	14,000 ft. density altitude
Cruise speed/75% power	113 kt. IAS
Max. range, no reserve	400 mi.
Hover ceiling, IGE, 2,400 lb.	6,100 ft.
Hover ceiling, OGE, 2,200 lb.	4,900 ft.

All-composite Boeing 360 blends speed, automation

Nicholas C. Kernstock/Wilmington, Del.
November 5, 1990

The Boeing Model 360 Advanced Technology Helicopter combines a low-drag composite fuselage, high-speed airfoils and integrated cockpit electronics into one of the most advanced rotary-wing aircraft ever built. The aircraft was conceived in the early 1980s as a technology testbed and also was used to investigate aerodynamics, vibration control, flight control and avionics systems.

Boeing has applied to other helicopters much of the technology developed for the 360, such as the stub wings and composite rotor blades on the CH-46, the glass cockpit on the MH-47E and a number of structures on the Boeing/Sikorsky LH contender. The company now is considering an attempt to set a new helicopter world speed record with the 360.

This *Aviation Week & Space Technology* pilot recently flew the Model 360 demonstrator at Boeing Helicopters' flight test center near Wilmington, Del. During the 90-min. flight, I was able to evaluate the aircraft's performance and handling characteristics as well as many other technical elements. Frank H. Duke, Boeing flight test pilot and director of the Model 360 project, accompanied me on the flight.

Model 360 technology demonstrator aircraft uses low-drag composite construction and advanced airfoils to give it a demonstrated airspeed of 214 kt. and a maximum design speed of 235 kt. The current world speed record of 217.4 kt. is held by a Westland Lynx.

The 360, the world's largest all-composite helicopter, first flew on June 10, 1987 (AW&ST July 13, 1987, p. 143). All of the primary and secondary structural elements of the fuselage, all of the structural attachment fittings and most of the dynamic components such as rotor blades, hubs, swash plates, shafting and rotor controls are made from composite materials.

According to Boeing engineers, the primary objective of the project was to develop design and manufacturing technology. This technology was to incorporate improved damage tolerance, lower construction costs than metal structures and, most importantly, substantial improvements in payload and speed. The result was a 20% increase in payload, a 30% speed improvement and a 60% increase in productivity over the rotary-wing technology available at the time. The 360 has demonstrated true airspeeds of 214 kt. in level flight and has a maximum design speed of 235 kt. The current world record of 217.4 kt. is held by a Westland Lynx.

Following a brief discussion on the design and construction of the airframe, Duke described the flight control system and avionics suite. The aircraft is equipped with a dual automatic flight control system (AFCS) provided by a pair of Honeywell dual digital computers that stabilize the helicopter in all axes and automatically maintain desired airspeed, bank angle and heading. The AFCS computers operate on signals from sensors that drive hydraulic actuators located in the primary control paths.

The computers use Mil-Std-1750A single chip processors operating in higher order language JOVIAL-73 and process flight control laws at a 40-Hz. rate. An Allied Bendix 1553 data bus distributes data throughout the system. The system permits the tailoring of command responses and stability according to flight conditions. A cockpit-mounted AFCS control panel allows the pilots to make gain changes using the keyboard to optimize the handling qualities in flight.

The system is a rate command type in pitch, roll and yaw axes. This means that a left cyclic control input commands a left roll rate rather than a rotor system position. To stabilize the rate, the pilot would center the cyclic control. To level the aircraft, the pilot would move the cyclic to the right and then center the control when level. Most conventional, nonelectronic flight controls use position sensing in which a left cyclic input would command a proportional amount of left bank. The pilot would level the aircraft by centering the cyclic. The collective pitch in the 360 uses conventional position sensing.

The 360 also has a three-axis coupled flight director that permits hands-off flying and a variety of automatic modes, including heading select; VOR and Doppler navigation; ILS localizer, glideslope and back course ILS; pressure and radar altitude hold; approach to hover and hover hold; and go-around and recovery.

Fine adjustments of the AFCS and flight director can be made using the "Chinese hat" trim switch on the cyclic and a "beep" switch on the collective control. Both AFCS computers incorporate built-in test circuits.

The 360 also features an all-glass cockpit consisting of a cockpit management system and four monochromatic cathode ray tubes with interchangeable electronic attitude display indicator (EADI) and horizontal situation indicator (EHSI) functions. The system consists of a pair of multifunction displays (MFD), two alphanumeric keyboards and dual mission computers.

The MFDs are 5-×-5-in. color CRTs that have a filter to allow night vision goggle compatibility. Each MFD has 25 push-button switches located around the outer edge of the display. The top five and bottom five switches are dedicated to specific functions, such as communications control, system status, pilot aids—such as checklists—AFCS gain control and performance, nav/com radio tuning, various system controls, and master caution acknowledge and reset. The switches on the sides control various functions available on the CRT pages.

After the briefing, Duke and I conducted a walk-around inspection of the aircraft. Immediately noticeable was the lack of any significant drag-inducing projections on the fuselage. The two Lycoming AL5512 engines are submerged completely within the aft rotor pylon. The configuration originally called for no engine intake openings in order to minimize flat plane drag. Engine air was ducted into the inlets from an opening at the top of the pylon. According to Duke, this arrangement resulted in higher engine operating temperatures. For purposes of the speed trials, the improved engine efficiency of the open intakes outweighed the drag penalty.

Seven primary frame members

The main landing gear support structures and retraction beams are constructed entirely of composite materials. The graphite fiber beams each weigh 45 lb. less than a similar steel design. When retracted, the gear doors fit flush with the fuselage.

Entry is made through the hydraulically operated rear ramp. The interior dimensions of the 360 are almost identical to the CH-46 but are much less cluttered. The use of composite construction allows significantly fewer frame and longeron sections. The exterior panels are made of Kevlar skins over Nomex honeycomb cores, and frames and longerons are fabricated from unidirectional graphite fibers in an epoxy matrix.

The airframe was built on seven primary frame members and, according to Boeing engineers, requires only one assembly fixture. Conventional metal aircraft construction often uses many more subassemblies and fixtures.

The cabin floor and fuel tank assembly was built as a module and is separated from the airframe by eight vibration isolation units. The cockpit is also modular in construction and is similarly protected from airframe vibration in all six axes. The cockpit is open and uncluttered, as is the instrument panel. Outside visibility is excellent even with the seat adjusted to the full down position.

An external APU provided electrical power to run the environmental control unit and instruments as Duke recapped the flight control and avionics briefing. Following an automatic start, the onboard APU powered the electrical systems, the hydraulic flight control actuators and main engine hydraulic starters. In flight, hydraulic power is provided by two mechanical pumps mounted on the forward and aft main rotor gearboxes and a utility system driven by the aft transmission. Main engine start is automatic, and systems checks were performed quickly as a result of the computerized data system.

The absence of rotor noise was very apparent as Duke moved the engine condition levers, located on the center console, to flight position and the rotors reached full operating rpm. The 360 uses a four-bladed, fully articulated rotor system incorporating a tapered planform and chord. Construction is similar to other composite blades on Boeing helicopters. The 360 uses a unidirectional glass fiber and graphite hybrid spar, Nomex honeycomb core and a high modulus graphite skin to provide torsional stiffness needed for high-speed flight. A nickel erosion strip is room-temperature bonded to the leading edge. The blades use a transonic airfoil (VR12/15) and increase hover efficiency by 6% and cruise efficiency by 23% over the CH-47D rotor system.

Duke wheel taxied from the ramp to the taxiway as I familiarized myself with the engine instrumentation on the multifunction display.

Boeing 360 has the agility for terrain flying and its glass/graphite-fiber hybrid rotor blades have the impact strength required for operations in trees. In service, these blades have cut down 4-in. trees without damage. During 200-hr. fatigue tests with saw cuts and 23-mm. high-explosive round hits in the spar, only a gradual softening and crazing was observed.

Main landing gear retraction beam is fabricated of graphite fiber composite. Each beam weighs 145 lb., a savings of about 45 lb. over a steel design.

The nose wheel is hydraulically steerable and electrically controlled by a knob on the console.

Duke pulled the 360 up to a 20-ft. hover and gave me the controls. Even with the auto-hover disengaged, the AFCS provided ample stability and hovering was relatively easy. Yaw response was surprisingly quick, considering the 51-ft. length of the aircraft. Landing and lift-off in a hover was smooth, and the aircraft exhibited no adverse roll sensitivity while light on the wheels. Position sensing switches on the main gear struts automatically reduce the pitch axis gain of the AFCS when the strut is compressed on touch down.

Cleared for takeoff

I made several pedal turns to either direction and soon was able to control the center of rotation along the longitudinal axis with little difficulty. The 360 displayed the tandem rotor configuration's characteristic insensitivity to wind direction and side-slip angle, and we hovered downwind with no noticeable difficulty.

After a few minutes, the tower cleared us for a takeoff from the taxiway to the south with a straight-out departure from the control area. I pulled a few percent of torque above hover power and lowered the aircraft nose to the horizon bar on the EADI, and the aircraft accelerated forward smoothly.

After confirming that we had established a climb, I looked at the airspeed indicator and saw it passing through 155 kt. on its way to 160 kt. before we had even cleared the airfield boundary fence. I reduced collective and pulled the nose up to hold 1,000 ft. MSL and 150 kt. airspeed until we had cleared the airport traffic area.

Duke took the controls and navigated to the Boeing flight test area, and I continued to operate the MFD and electronic flight instrument displays. The combination of clear function markings and CRT prompts made the glass cockpit fairly simple to use. The system can provide a wealth of information but allows the pilot to control how much is displayed.

Upon arrival at the test range, we conducted a reconnaissance of the nap-of-the-earth (NOE) area and waited for a photo/chase aircraft. Duke and I flew several runs at speeds from 50–90 kt. and altitudes from 5–30 ft. We approached a treeline from a masked position and executed a quick cyclic climb followed by a pushover into a small clearing to remain masked. The 360 maintained good control

Fully integrated electronic cockpit management system takes the place of about 40 round dial gauges and 150 switches found in helicopters of similar size.

response with no significant increase in vibration in the pushover and smoothly returned to full positive-g flight. During the flight test phase, the 360 attained bank angles of 60 deg. and pushovers and pullups from +.5 to +1.5 g.

Skirting the edge of the clearing required several sharp coordinated turns, and the agility of the 30,500-lb. aircraft was surprising. Control accuracy is an essential factor in NOE operations and, after several passes, I found I was able to consistently place the rotor tips within a foot of obstacles while at speed.

The final series of maneuvers called for another climb and pushover, followed by a sharp right descending turn to exit the course at an altitude of 5–10 ft. over a flat marsh. Here, the overhead cockpit visibility was important in clearing the turn around a group of trees.

We each performed several masking and unmasking maneuvers from a hover and, once again, the AFCS delivered precise control regardless of wind direction or mechanical turbulence.

NOE quick-stops were more challenging because the high pitch-up attitude required to stop the aircraft made it difficult to stay masked. Given enough room a sideward stop could be used, but we did not demonstrate the maneuver.

Following the low-altitude work, we climbed back to 1,500 ft. to evaluate some of the aircraft's yaw characteristics in cruise flight. Because of its low-drag design, the 360 does not naturally point into the incident wind. In single-rotor helicopters, the tail rotor and empennage create enough drag to stabilize the aircraft longitudinally. In modern tandem-rotor aircraft, the flight control system stabilizes the yaw axis electronically by providing control inputs to the rotors.

To demonstrate the 360's yaw stability, Duke disengaged the AFCS while I maintained straight and level flight. The antitorque pedal workload immediately increased and, while the aircraft was manageable, it required constant attention to maintain heading. The aircraft has flown at sideslip angles of ±15 deg., but sideslips in excess of 30 deg. in cruise flight could result in excessive side loads.

With the AFCS engaged, I executed several turns with up to 45 deg. of bank to investigate its limits for yaw stabilization in turns. As in most tandem-rotor helicopters, the 360's aft rotor system can stall in high-g turns and cause an abrupt out-of-trim condition. Upon returning to the airport, I wheel taxied to the Boeing ramp for shutdown.

According to Duke, one of the major successes of the 360 project was the reduction of costs realized by the use of composite construction. In a comparison of man-hour costs of prototype tooling, the 360 cost $1,153 per square foot versus $15,000 per square foot for a conventional sheetmetal airframe. Boeing engineers expect similar reductions could be achieved for production aircraft because of the opportunities for automation that composites offer.

Duke and other Boeing engineering and marketing personnel also maintain that the Model 360 is not an alternative for the Bell/Boeing V-22 Osprey. According to one company official, if the V-22 is killed in the Defense Dept. budget, the 360 is unlikely to be considered as a replacement because it would require an entire new program proposal. Given the time and expense involved, a restart of the CH-46 production line probably would be preferable.

Interior dimensions of the 360 are nearly identical to the Boeing CH-46's. The 360 can carry an internal payload of 14,000 lb., and the main cargo hook support structures are designed to carry an external load of 30,000 lb.

Boeing/Sikorsky Fantail allows radical maneuvers

Nicholas C. Kernstock/West Palm Beach, Fla.
August 27, 1990

The Boeing/Sikorsky H-76B Fantail demonstrator aircraft, equipped with the companies' fan-in-fin antitorque system, delivers unrestricted yaw pointing capability at airspeeds up to 80 kt. as well as other unusual agility and maneuverability characteristics. The Fantail was developed jointly by Sikorsky Aircraft and Boeing Helicopters as part of their Light Helicopter design.

This *Aviation Week & Space Technology* pilot flew the H-76B Fantail prototype at Sikorsky's West Palm Beach, Fla., flight test facility. Accompanying me in the right seat was Sikorsky chief test pilot Nick Lappos. During the 1-hr. flight, I was able to evaluate the antitorque system's performance in a variety of maneuvers, including many specified in the LH program requirements guide.

The parent aircraft is an S-76B that has been configured for military operations. Following a briefing on the maneuvers and data points we expected on the flight, we conducted a quick preflight inspection of the Fantail helicopter. After engine start, the Fantail's low noise level was noticeable. Also absent was the medium-frequency vibration associated with conventional tail rotors.

Boeing/Sikorsky Fantail.

As I wheel-taxied from the ramp area to the active runway, I made several turns using antitorque pedals and differential braking. When cleared onto the active runway, I brought the aircraft to a 5 ft. hover in a light, right crosswind. With the stability augmentation system (SAS) engaged and control position trim off, the aircraft's handling was docile. After I executed several hovering turns to reacquaint myself with the H-76, Lappos took the controls to demonstrate several maximum-rate pedal turns.

180-deg. turns in 4 sec.

LH program requirements call for 180 and 360 deg. in-ground-effect (IGE) hover turns to be accomplished while maintaining the axis of turn within a 1.5-meter circle at 0 airspeed and a 3-meter circle at 45 kt. The maximum vertical travel cannot exceed ±2 ft. and the heading tolerance at stop ±2 deg.

I used a stopwatch to time the turns as Lappos executed a right 180-deg. turn in 3.75 sec. and another to the left in 4.23 sec. This was followed by a light 360-deg. turn in 5.05 sec. and one in 5.35 sec. to the left. Throughout the maneuvers, our position over the ground remained within a 1-meter circle, vertical position was within ±1 ft. and final heading was within the LH requirement. The maximum yaw rate attained was about 100 deg./sec. I took the controls and made several turns at less than maximum rate, speeding them up gradually. Executing the maneuvers was straightforward.

I hovered the aircraft at 8 ft. and centered the cyclic control. As Lappos counted down to zero, I pressed the right antitorque pedal as hard as I could, counted to three and pressed the left pedal to the control stop. The aircraft spun to the right instantly. As the nose swung around to the runway centerline, I neutralized the pedals and the rotation stopped. The maneuver was very abrupt but remained controllable.

After several practice turns, I was able to turn 180 deg. in about 4.05 sec. and 4.30 sec. to the right and left, respectively, and 360 deg. in about 5.25 sec. and 5.50 sec. The tendency to overcontrol the end of the turn was minimal, and I was able to stop within ±5 deg. of desired heading.

One of the most noticeable characteristics of the Fantail was the lack of roll/yaw coupling. The tendency of an aircraft to roll to one side when large yaw control inputs are made is the result of positioning the tail rotor above the aircraft's horizontal axis. By mounting the fan on the horizontal axis, roll/yaw coupling is almost eliminated.

The Army's LH sideslip angle or crosswind requirements are significantly higher than for current-generation helicopters. The maximum crosswind for the Bell UH-1, AH-1 and Hughes OH-6 was 17 kt. UH-60 and AH-64 helicopters have a maximum allowable sideslip of 35 kt. The LH design criteria, by comparison, require the aircraft to make 180 deg. turns in a wind of 45 kt.

Lappos demonstrated sideward flight of 70 kt. to the right, 65 kt. to the left and rearward flight at 70 kt. For sideward flight, the aircraft hovered at 90 deg. from the runway centerline and accelerated to the side. The accelerative attitude was maintained until we could no longer hold the 90 deg. orientation. I found the maneuvers easily manageable, although left sideward flight at 65 kt. from the left seat resulted in an uncomfortably low seating position.

New challenges

The most impressive capability was the aircraft's unlimited yaw pointing ability at speeds up to 80 kt. Flying down the runway at about 20 ft. altitude, I applied full left pedal and adjusted the cyclic control to

The vertical strut supports the Fantail's gearbox. By placing it off the axis of fan rotation, shock waves created by the blades are suppressed. The horizontal strut houses the drive shaft.

maintain ground track. I was able to point the nose of the aircraft in any direction at will.

Following our maneuvers on the runway, we climbed to 1,500 ft. to evaluate the yaw pointing characteristics in out-of-ground-effect flight. One of the LH maneuvers we performed was the coordinated 90 deg. turn to target at a constant airspeed and altitude. LH requirements specify the turn be completed in less than 6 sec. This maneuver would be used to quickly bring the aircraft into firing position after identifying a threat at its 90 deg. position. The conventional maneuver is a steeply banked turn to target, and we were able to execute this turn in 3.32 sec. However, using the Fantail's unrestricted yaw capability, we made a turn using only the antitorque pedals in 2.28 sec. This 1-sec. advantage would be of significant value in an air-to-air combat situation.

I also executed pedal turns while maintaining constant ground track. I was able to initiate the turn at 120 kt. and hold a stable pointing attitude as the aircraft slowed to about 75 kt. Allowing 10 sec. to target, fire and track a TOW or Hellfire missile and take evasive actions, the entire maneuver could be completed in less than 20 sec.

Other maneuvers included 180-deg. "snap turns" in which the flight path is reversed by making a pedal turn in much less time and distance than a coordinated turn. In several high-g accelerating turns and pull-ups at 130–150 kt., the aircraft showed no adverse yaw characteristics.

Sikorsky test pilot Nick Lappos has demonstrated sideward flight in the H-76B to the right at 70 kt. and left sideward flight at 68 kt. Airspeeds of 70 kt. have been attained in rearward flight.

Fan blades can be individually removed for maintenance through the plug in the shroud, visible at the bottom left of the duct. The graphite and glass fiber blades have an infinite fatigue life.

The Fantail far exceeds the abilities of conventional tail rotors in several respects, and many of the maneuvers the system is capable of do not yet exist in combat flying doctrine. The selection of a Fantail-equipped LH would provide Army and Boeing/Sikorsky test pilots with some unique and interesting challenges.

Armed S-76 helicopter provides low vibration, good stability

David A. Brown/Farnborough, England
November 22, 1982

Sikorsky S-76 Mk. 2 Plus, armed version of the utility and light transport helicopter being produced by Sikorsky Aircraft Div. of United Technologies Corp., is a responsive and maneuverable helicopter that benefits from low vibration levels, which give it good gun-platform stability.

Flown here by this *Aviation Week & Space Technology* pilot, the Armed S-76 exhibited:

- High maneuverability in nap-of-the-earth flying.
- High cruise speed with external weapons mounted.
- Ample control forces in both sideward and rearward flight.
- Good rates of climb with one engine out.
- High stability level with stability augmentation systems inoperative.

The Armed S-76 concept was derived from earlier efforts to expand the market for the basic S-76, which initially was developed

Sikorsky S-76 Mk.2 Plus armed helicopter makes a rocket firing run during development testing. The helicopter can carry a variety of weapons, including 2.75-in. folding fin rockets (shown in the photo), 7.62-mm. and .50-cal. machine guns, both forward-firing and side-mounted, forward-firing 20-mm. guns, and up to eight Hughes TOW missiles.

with offshore oil rig support as the primary goal. The design requirements for the offshore support role called for a 400-naut. mi. radius of operation carrying 12 passengers. To achieve this range, the original design stressed a low-drag fuselage.

The S-76 corporate version was added to the line because the helicopter had a cabin large enough to make it an effective transporter of management teams, the designers believed. Following the corporate version, a utility version was developed, and the two types are basically interchangeable.

The Armed S-76 is a greater modification of the original helicopter than any of the previous derivatives but retains the basic S-76 airframe, power train and engines, although new engines are still being discussed and a decision is likely soon (AW&ST Sept. 13, p. 24).

Farnborough site

I flew the Armed S-76 from the left seat with Sikorsky test pilot Edmund K. Dinsmore in the right seat. The helicopter was operated from a small, private helicopter pad near Farnborough, site of the Royal Aircraft Establishment's research facility.

The helicopter, with U.S. registration N5415X, was equipped with two Belgian-made Fabrique Nationale TMP machine-gun pods. One is mounted on each of the two weapons pylons, which in turn are

Cockpit of the Armed S-76 is the same as the civil version of the aircraft, with the exception of the weapons control on the cyclic stick and a gunsight atop the instrument panel.

mounted on the helicopter via a boom that passes through the aft cabin of the Armed S-76 and is bolted to the airframe.

Takeoff weight of the aircraft was approximately 8,500 lb., or about 1,000 lb. below maximum gross liftoff weight. This weight included the two machine-gun pods and the boom that supported them. A mast-mounted sight for the TOW missiles was not carried on this flight.

Dinsmore started the two Detroit Diesel Allison 250-C30S turboshaft engines after a brief demonstration of the cockpit. Cockpit layout was the same as the civil version in this instance, with the exception of the addition of a gunsight mounted on the right-hand glare shield. The engine instruments were grouped on the right of the central display panel, with duplicate flight instruments located in front of each pilot. Communications and electrical switches were grouped on a pedestal console between the two pilots, and systems operating switches, trim controls and radio transmission buttons were on the cyclic and collective controls.

After releasing the rotor brake and running a quick controls check, Dinsmore invited me to lift the Armed S-76 into a hover. First impression after liftoff was of the very light control forces needed to hold the helicopter in one place. These are lighter in the Armed S-76 than in many other similar helicopters. Although this is a matter of individual pilot preference, it was something that I at first disliked but became adjusted to after a few minutes' flying, and from that point on I found it useful.

The helipad from which we were operating was surrounded by tall trees, which prevented a normal accelerating takeoff. Dinsmore took the opportunity to demonstrate the S-76's vertical liftoff capability by performing a vertical climb to 150 ft., then accelerating to single-engine speed. The maneuver has been approved by the British Civil Aviation Authority for use by charter operators using the helipad.

We departed the Farnborough area for the abandoned airport at Wisley, England, which once served as the assembly site for Vickers Viscount aircraft but which now has only a hard-surface runway, and all the buildings have been demolished. The airfield provided a clear space for low-level work, and the area immediately around the runway had some undulations and trees to permit nap-of-the-earth demonstration flying.

En route, Dinsmore demonstrated the Armed S-76's speed capability, and I flew some gentle turns to get the feel of the control system in flight. Never-exceed speed (V_{ne}) with weapons pods on is listed as 155 kt., although Sikorsky believes this can be increased later in the development program.

At 100% power, we were able to stabilize the Armed S-76 at 141 kt. In level flight, cruise speed is listed at a nominal 110–115 kt. Outside air temperature at this time was 20C.

We descended to a hover at Wisley and made several low-level passes down the runway, followed by sideways and rearward flight. Flight to the right was simple, with the helicopter easily accelerating to 35 kt. Flight to the left is a bit more difficult, since the control system power limit is reached at about 16 kt. Once the helicopter has accelerated past this speed, however, it acquires enough translation lift to ease the control power requirement, and I was able to get it up to 40 kt.

Rearward flight

Flight to the rear required a little work on the rudder pedals, but the Armed S-76 would hold 30 kt. without difficulty. In hovering flight, in rearward flight close to the ground or in the approach to a hover, the pilot has to take care not to get too high a nose angle close to the ground, since at only 10-deg. nose-up the tail boom is level and there is no skid to prevent dragging the tail on the ground. The tail rotor sits high enough that it would not drag the ground until a higher angle was reached.

Since the S-76 has a relatively low aerodynamic drag, even with gun pods on, it tends to decelerate slowly. The result of these two factors is that it is necessary to start decelerating early, by getting the nose up at a relatively high altitude, to make the hover at the desired spot.

We were hovering and Dinsmore was discussing nap-of-the-earth techniques used when flying the Armed S-76, prior to a demonstration of the helicopter's capability, when our flight was interrupted briefly by a British bobby from the Surrey County Constabulary who rode up on his bicycle and flagged us down. He inquired if we had permission to use the airfield, took the aircraft registration number and, after a few pleasantries, allowed us to continue.

During the nap-of-the-earth flying, the Armed S-76's sensitive controls are demonstrated to their best advantage. The flight control system is the same, except for the physical layout, as that in the larger Sikorsky UH-60 Black Hawk helicopter.

It incorporates dual independent hydraulic systems. Control servos have dual cylinders arranged in parallel and with a separate hydraulic pump driving each cylinder. The tail rotor has the dual cylinders arranged in-line for ease of maintenance.

All of the servos are of a jam-proof design, and each normally operates at only 50% capacity, so that in case of servo failure, the second one will have the same power as is normally provided by both. Control loads are not noticeably changed by a servo failure, Dinsmore said.

Detachable weapons pod system is shown attached to the S-76 Mk.2 Plus and carrying a 7.62-mm. gun pod.

Detachable weapons pod system detached and on display at this year's Farnborough air show, in front of the helicopter. Two attach points bolt to the cabin floor aft of the pilot's seats.

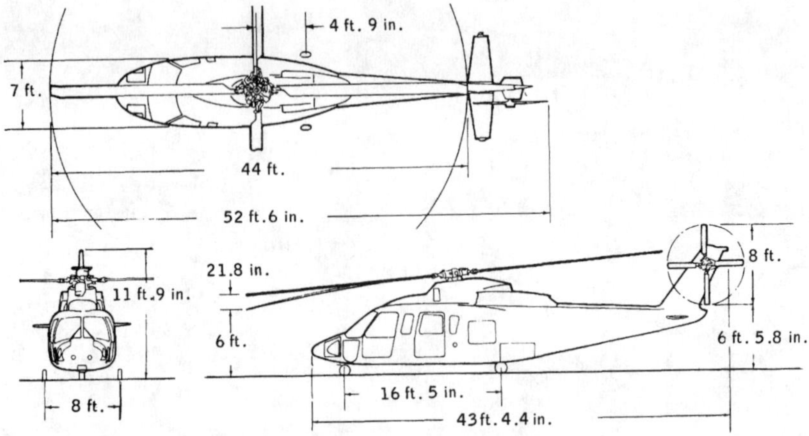

S-76 rotor tilts 5 deg., permitting pods to be mounted at 0-deg. angle of incidence for forward firing. Rocket pods are mounted at +5 deg.

Also contributing to the helicopter's performance in nap-of-the earth flying are the main and tail rotor designs. Both are based on the National Aeronautics and Space Administration 1095 airfoil, with the main rotor airfoil modified by Sikorsky to have an 80% drooped leading edge inboard for improved lift in hover and low-speed flying. The main rotor blades also have an 8% pretwist for better low-speed flight.

The main rotor blades have a hollow titanium spar and leading edge erosion strip with a glass fiber skin over a Nomex honeycomb core. The rotor blade tips are swept, which reduces the noise generated by the blade and also produces an aerodynamic force that reduces the built-in blade twist at higher speeds to improve retreating blade stall characteristics.

Problems corrected

The spindle problems that plagued earlier versions of the S-76 have been corrected, Sikorsky officials said, and increased spindle lifetimes have been approved for the civil version of the helicopter by the U.S. Federal Aviation Administration. The spindle and hub assembly used on the Armed S-76 is a scaled-down version of the system used on the UH-60. It is made of forged aluminum with two elastomeric bearings. It requires no lubrication and has no roller bearings.

The tail rotor also is a spinoff from the Black Hawk and is a soft, in-plane rigid rotor with no bearings. In case of a hydraulic failure, the tail rotor automatically assumes a 6–7-deg. positive pitch angle, which permits a safe power-on emergency landing to be performed.

Dinsmore demonstrated the Armed S76's nap-of-the-earth capability with a flight across the area adjacent to the runway at Wisley, which is cut by old roads, the foundations of old buildings and trees.

One factor that became noticeable at this time was the relatively good visibility afforded by the cockpit design, even though it originally was for civil rather than military use.

A redesigned panel and glare shield, which would be a probable change if the Armed S-76 were ordered into production, would possibly improve vision directly forward and down, but lateral visibility and visibility downward to the side is excellent.

Another factor evident in the low-level operation of the Armed S-76 is the low noise level from the rotors. At times the only noticeable sound in the cockpit was the transmission noise.

The Armed S-76's light control forces, particularly the collective control, were displayed to good advantage during low-level flight and particularly during simulated unmasking operations, where it was simple to pop up to a desired height, stabilize quickly and return to a lower hover with a minimum of effort.

Since it is likely that the Armed S-76 would operate largely at low altitudes in a military environment, a number of modifications were made to the basic airframe to enhance safety for the crew in the event of a crash. Much of this work was based on experience gained in Vietnam. Major change was in the aircraft's fuel system, which has an engine-mounted suction fuel pump rather than the pressure pump used previously.

Experience has shown that if there is a break in the fuel system at any point, a pressure pump will force fuel through it, possibly onto hot areas, which could result in a fire. A suction pump will not, however, since it effectively is shut down as soon as the fuel system integrity is destroyed. Crashworthy fuel cells also have been added.

Mountings reinforced

Both the transmission and engine mountings were reinforced to hold the engines and transmission in place in the event of a crash landing, and the aircraft tub—the bottom portion of the fuselage—was insulated with rails or skid strips to assist in retaining cabin integrity in a crash.

After completing low-level work at Wisley, we took off and climbed to altitude to fly the helicopter with various systems inoperative or degraded and also to see the Armed S-76's tactical maneuvering capability. At altitude, Dinsmore shut down one engine while I was flying the helicopter to demonstrate that—aside from a small decrease in speed—there was no noticeable change in the Armed S-76's handling qualities or feel.

He then placed the helicopter in a 500-fpm descent. From this descending flight, he brought the aircraft up into a single-engine climb, which stabilized at just under 500 fpm with the good engine pulling 111% of torque and the rotor turning at 96%. Twin-engine climb had stabilized at 850 fpm at an engine turbine temperature setting of 828C above the maximum continuous rating of 796C.

System capability

With one engine inoperative, Dinsmore continued to degrade the helicopter's systems capability by securing one of the stability augmentation channels, then an alternating current generator and finally a control servo.

Other than a slight bump in the control feel when the helicopter switched to inverter power after the failure of the a.c. generator, there was no indication from the helicopter's flying qualities that there had been any loss of performance.

Final demonstration was the Armed S-76's tactical maneuvering capabilities. These included a tactical turn with an angle of bank of 115 deg. The Armed S-76 has performed Split-S maneuvers during the test program, although we did not attempt one on this flight because of the presence of a third crewman in the rear.

Dinsmore also demonstrated the Armed S-76's ability to perform less than 1g maneuvers and, concurrently, the stability it can provide as a weapons aiming platform, by accelerating to about 140 kt., then pulling up to an 18-deg. nose-up angle and flying the Armed S-76 over the top of an arc. The g load was estimated at about 0.2 of normal, and while the helicopter was flying at less than 1g, I rolled it sharply left and right to determine control capability, which was positive and ample.

During the maneuver I held the nose of the helicopter on a prominent landmark, in this case a building, and tried to assess the vibratory movement. At speeds up to the maximum we reached—which was just under 160 kt. indicated airspeed—the vibration levels were light and noticeable only at the highest end of the speed spectrum. They never interfered with aiming or holding the helicopter precisely on a point.

Sikorsky has engineered a stickshaker into the control system, which vibrates the cyclic control any time the Armed S-76 is flown into a maneuver where loads on the rotor control system reach 80% of the system's endurance loads. When the vibration occurs, the pilot generally can eliminate the condition by lowering the collective slightly and unloading the entire rotor system. If a tactical situation were to prevent this, the pilot at least would know he was operating on the edge of the rotor's control capability range.

Since the Armed S-76 is a relatively high-speed, stable platform, Sikorsky has moved rapidly to qualify a number of weapons for use with it. These include:

- The SUU-11 door-mounted Gatling gun system.
- The Belgian TMP and HMP gun pods.
- The GPU-2A 2.75-in. rocket pod, which carries 19 rockets per pod.
- The Hughes TOW wire-guided antitank missile.

Other weapons systems are expected to be qualified in the future.

Modification made in the course of adding the weapons capability to the helicopter included adding a cargo floor to the cabin capable of accepting a 200-psf. load and additional under-floor structure to support eight mounting points to which various armament systems can be attached.

Electrical winch

Electrical connections also were added for weapons control, with an umbilical cable located in the helicopter fuselage tub. At present there is a separate cable required for each different weapon, but in a production design these would be unified.

A 600-lb. capacity winch has been added that can be operated with the electrical power from one of the aircraft's d.c. generators. A Saab-Scania RSG-1 gunsight has been tested with the helicopter, although it has not yet been qualified for use.

We concluded our flight with a return to Farnborough and a landing into a relatively tight, confined space, bordered by trees on two sides, a tall fence on a third, and with another helicopter on the adjacent pad. Again, the helicopter's good visibility and easy hovering maneuverability made what could have been a difficult approach seem relatively simple.

Sikorsky S-76 stresses performance

Robert R. Ropelewski/Las Vegas
February 19, 1979

Stability, performance, vibration and interior noise levels equivalent to fixed-wing aircraft are evident in the Sikorsky S-76 utility and corporate helicopter that is entering service with Air Logistics, Inc., this month. Sikorsky's first commercial delivery of an S-76 took place early in February in Lafayette, La., where Air Logistics, a subsidiary of Offshore Logistics, Inc., is headquartered. The aircraft, the first of 14 ordered by Air Logistics, will be used to support oil drilling operations in the Gulf of Mexico.

Delivery of the first Air Logistics S-76 was preceded by several days by Federal Aviation Administration Category A and instrument flight rules (IFR) type certification of the aircraft. The basic type certification for the S-76 (Category B and visual flight rules) was issued last November. Sikorsky officials believe the Category A and IFR endorsements make the S-76 the first transport category helicopter designed specifically for the commercial market to carry the FAR 29 Category A and IFR certification from its initial entry into service. The

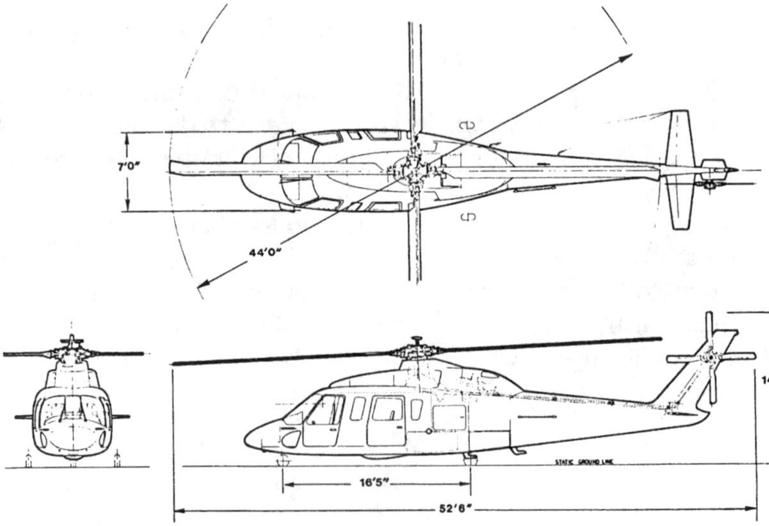

Three-view drawing of the Sikorsky S-76 shows general dimensions of the utility/executive helicopter as it will go into commercial service.

Category A certification means the S-76 at its maximum gross weight of 10,000 lb. can continue its mission after losing one engine up to a temperature of 90F and 1,000 ft. altitude, and still maintain a climb rate of at least 150 ft./min. if necessary.

This *Aviation Week & Space Technology* editor flew the S-76 on a 40-min. evaluation flight during the Helicopter Assn. of America 31st annual convention and industry showcase in Las Vegas. The stable handling characteristics, low vibration and low interior noise levels observed during this flight and an earlier flight as a passenger underscored progress being made in advancing helicopter technology to match fixed-wing standards.

The aircraft flown was the fifth S-76 to be built, and the first production aircraft. It will remain a service evaluation aircraft for Sikorsky, however, and will not be delivered to a customer. The aircraft was configured for utility operations, like offshore oil drilling support, with seats for two pilots and 12 passengers.

In this version, the passenger cabin consists of three rows of four-abreast seating, with 29.5-in. pitch between rows, and seats that Sikorsky officials said were 1 in. wider than those found in the tourist section of the average Boeing 727 transport aircraft. This is approximately the configuration that was to be delivered to Air Logistics.

A plushier, six-passenger executive transport version of the S-76 also was exhibited at Las Vegas, although this aircraft was not demonstrated in flight. Sikorsky is offering customers six floor plans in the executive configuration, accommodating from four to eight passengers. In addition to the greater space available in the passenger cabin, the executive version also offers a higher degree of soundproofing than is available in the utility configuration.

The evaluation flight in the utility version S-76 was made from North Las Vegas airport with Sikorsky test pilot Richard Brosius after an extended preflight briefing by Sikorsky engineering test pilot Richard L. Mills, who has been on the S-76 program since its initiation. The aircraft weighed approximately 7,700 lb. at the start of the evaluation, including Brosius, this editor, a passenger and just under a full load of fuel. Two other passengers were taken aboard later in the flight evaluation.

Empty weight of the S-76 with standard equipment is advertised as 5,273 lb., and the maximum gross weight is 10,000 lb. Fuel capacity is about 280 U.S. gallons, or approximately 1,820 lb. The No. 5 aircraft used for the evaluation had a total of about 150 flight hours at the start of the flight.

The S-76 is equipped with a door for each pilot and one door on each side of the passenger cabin. Door handles and locks are stan-

Offshore interior configuration of the No. 5 aircraft has three rows of bench seats that are 75 in. wide. Seat pitch is 31 in., measured from seat centers, and there are 15.5 in. between seats in the leg area. A 6-ft.-tall passenger can sit without undue crowding in the two rear rows, but in the center two seats of the first row, his knees will touch the backs of the pilot seats. Each four-unit bench is removable from the aircraft.

dard automotive equipment, designed to be operated by non-aviation-oriented personnel. Lock pins, for example, are identical to those used on automobiles. Separate handles are provided for emergency door jettison.

Entry to the two rear rows of seats in the passenger cabin is accomplished easily through either doorway to the cabin. Entry to the pilots' compartment and to the first row of seats in the passenger cabin is made through the forward door on either side and involves a slight squeeze for passengers going into the first row of cabin seats. This is because of the limited space through which passengers must pass between the doorpost and the back of the pilots' seats.

The S-76 has been certificated for dual-pilot instrument flight rules operation, and this particular aircraft was configured for IFR flight from either side of the cockpit. Instrument panel on each side

Executive interior of the No. 6 aircraft, which will go to Sikorsky's parent, United Technologies Corp., has a rear 75-in. bench seat for three with armrests folded down or four with the armrests folded back. Two fully articulating chairs can recline 45 deg., rotate through 360 deg. and track forward and aft 4 in. and 2.5 in. inboard and outboard. Floor covering in this helicopter is 100% wool deep-cut pile. Noise level maximum of the executive version of 75 db. is 5 db. lower than the utility version of the aircraft.

of the cockpit was equipped with a 5-in. Sperry attitude gyro with a horizontal situation indicator below and with other primary flight instruments to either side of these displays in what has become the standard IFR arrangement on most aircraft today.

Circuit breaker panel between the backs of the pilot's and copilot's seats and intercom/radio selector boxes at the center of the front instrument panel will be familiar to veteran Sikorsky H-34/S-58 pilots, as will the location of stability augmentation and automatic trim system switches at the rear of the center console. Engine start switches and engine control levers are positioned overhead between the two pilots.

Standard prestart checklist is used, and the twin Allison 250-C30 turboshaft engines are started individually simply by pressing the engine start button on the engine control lever, then advancing the lever

First production version of the Sikorsky S-76 utility/corporate helicopter—the No. 5 aircraft off the assembly line—has a vertical stabilizer that is 12 in. shorter than the prototype. Other production modifications include nose section windows added to those on the fuselage sides below the cockpit and relocation of the pilot tubes to the nose section from their placement above the cockpit on the prototype.

to idle when gas generator speed reaches 12%. Engine temperature can be controlled during the start by moving the engine lever as needed between the shutoff and idle positions.

650-shp. engines

The engines, which produce 650 shp. each at takeoff, can be started with the rotor brake on or off. During extended waiting periods on the ground, this feature allows the pilot to operate heating or cooling systems without having to keep the rotors turning. Starting one engine with the rotorbreak engaged, then releasing it as the second engine is started, results in a smoother rotor turn-up with less rocking inside the aircraft.

Sikorsky test pilots had high praise for the Allison 250-C30 engines, which, they said, have performed very well despite the deliberately heavy abuse they have had during ground and flight testing. In two years of flight tests, only one inflight engine shutdown has been necessary. That occurred when an engine chip light illuminated in flight.

Engine levers (light knobs) are located on the overhead panel, and the engine start buttons in the knobs. Emergency fuel/oil shutoff handles (dark knobs) are also located on overhead panel.

Taxiing simplified

Taxiing the S-76 has been simplified by the aircraft's tricycle landing gear arrangement and by the fact that the rotor mast, and thus the rotor disk, is tilted forward 5 deg. Taxiing can be initiated with only a small amount of power, and without having to push the cyclic stick forward.

Once rolling, the aircraft can be steered easily with the rudder pedals. Several 180 deg. turns on the narrow taxiways and runway of North Las Vegas were no problem at all, and the aircraft could be pivoted around a main gear without difficulty.

With the aircraft on the runway and the outside air temperature reading 5C (41F), the aircraft's two stability augmentation systems were turned on and the S-76 was lifted into a hover. Field altitude was 2,200 ft. and a 6-mph wind was blowing from about 30 deg. left.

With the SAS on, a hands-off hover was established about 10 ft. above the runway. About 50% torque was required to hold the hover under these conditions. Full 650 shp. per engine was available at that temperature and altitude. Performance charts for the S-76 indicate the aircraft could have hovered in ground effect at its maximum gross weight up to a temperature of about 85F. Alternatively, at our weight of 7,700 lb. and the prevailing temperature conditions, it should have been possible to hover in ground effect up to 14,000 and out of ground effect up to 12,000.

Still in the hover, stability augmentation systems were turned off briefly. The S-76 continued to exhibit good stability characteristics, although closer attention had to be paid to aircraft attitude and drift to maintain a steady hover.

Stability augmentation was then reengaged, the nose was lowered slightly, more power was pulled in with the collective and transition was begun to cruise speed. The transition was exceedingly smooth, with little trace of vibration during the passage through translational lift.

Sikorsky engineers attribute the smoothness of the S-76 in the transitions from hover to forward flight and from forward flight to a hover to the swept blade tips of the aircraft, which they say soften the lift and drag variations acting on the rotor in these situations.

Rapid acceleration to a cruise speed of 145 kt. was also noteworthy during this portion of the flight, although the absence of a payload and the relatively light weight of the aircraft were prime factors behind this. Best rate of climb under the prevailing conditions was achieved at 72 kt. airspeed, where the climb rate stabilized at about 1,600 ft./min. The aircraft was accelerated to its normal cruise speed of 145 kt. at 3,500 ft., using about 55% torque per engine to maintain this speed. The S-76 was exceptionally smooth and quiet in this condition, as it had been during the previous hover and climb.

Power was added again and the speed was increased to 155 kt., the maximum allowable airspeed (V_{NE} for the S-76. About 92% torque per engine was required to maintain this speed. Vibrations increased noticeably between 145 and 155 kt., although instruments still were clearly readable and the ride was not uncomfortable. Current margins are such that Sikorsky engineers believe it will be possible to extend the S-76's maximum sea level speed to 165 kt. in the near future.

Brosius said the vibration level was considerably lower at higher weights, as when passengers are aboard the aircraft. On a later portion of the flight, three passengers were carried in the rear and rotor-induced vibrations disappeared almost completely.

Cruising at 155 kt., Brosius pulled the No. 1 throttle back to idle to simulate an engine failure. The aircraft yawed left very slightly before the stabilization systems caught up with the drop in torque, but hands-off flight was still possible. Collective lever was adjusted to set the No. 2 engine at 104% torque and 768-deg. turbine outlet temperature (T_5), and speed stabilized at about 130 kt.

Maximum power available was 111.2% torque and 826C T_5. However, this temperature setting can only be used for 2.5 min. The 768-deg. setting is both the normal takeoff temperature and the maximum continuous temperature limit for the 250-C30 engine. Other pertinent temperature settings for the engine include the 694-deg. engine warranty setting, 798 deg. for a maximum of 30 min. continuous running, and an 871-deg. transient limit.

The engine warranty setting was introduced by Detroit Diesel Allison for warranty purposes. While the 768-deg, maximum continuous temperature limit is applicable to FAA certification criteria, Allison's warranty on the 250-C30 engines is good only up to a maximum continuous operating temperature of 694 deg.

With both engines operating normally once again, landing gear was lowered at a speed of 130 kt. No noticeable trim changes were evident, and the aircraft continued to fly hands-off. Same stability was displayed when the landing gear was raised.

Sikorsky has included a mechanical linkage in the S-76 rotor system that decreases pitch when the blades begin to flap in turbulence. As a result, the aircraft shows a strong immunity to normal and moderate turbulence. The ride was exceptionally smooth during the demonstration flight, despite gusty, stormy weather that was generating substantial turbulence in the mountains north of Las Vegas where we were flying.

The smoothness was all the more impressive because this particular aircraft did not have all the fixes that will be incorporated in succeeding production aircraft. This includes a self-tuning damper located in the nose and connected to the rotor rpm sensor. Weights on a spring are moved farther apart or closer together to absorb rotor vibrations as rotor speed changes.

The No. 5 aircraft flown on this evaluation flight has what Sikorsky technicians refer to as a Phase 2 stability augmentation system, which is a rate-sensing system providing limited attitude hold in pitch roll and yaw. There is no feedback to the cycle stick or the rudder pedals in these cases.

Phase 3 SAS to be incorporated on late aircraft will include the Phase 2 features plus an outer loop in pitch and roll that maintains attitude, provides coordinated turns, holds heading and moves the stick to a neutral trim position whenever automatic control movements are made. Adjusting stick position in this manner allows the automatic system to maintain its authority through full stick throw. A Phase 4 AFCS is being studied that would provide airspeed and altitude hold, and could be coupled to an instrument landing system (ILS) to provide a hands-off instrument approach capability.

The S-76 also utilizes what Sikorsky engineers claim to be a jam-proof servo system that automatically isolates a jammed actuator anywhere in the control system, bypassing the actuator and allowing the rest of the system to operate normally. Redundant actuators for each control function ensure controllability in such cases. A warning light in the cockpit advises the pilot when an actuator has been isolated.

Turning off one of the two SAS systems or one of the two hydraulic systems in flight brought no noticeable changes to the flying qualities of the S-76. Turning off both SAS systems demanded more attentiveness in flying the aircraft, although basic stability remained good and the increased workload was still within comfortable limits.

Two approaches and landings made at North Las Vegas field—one with three persons aboard and one with five aboard, including pilots—were smooth and uncomplicated. The exceptional smoothness and stability of the S-76 is apparent throughout the flight regime, including approach and landing operations. The aircraft maintained a fixed attitude and decelerated smoothly on the approach with almost no effort on the part of this pilot. Yawing was almost imperceptible during power changes.

Vibrations were negligible during transitions to a hover, despite landing with a tail wind, and the aircraft consistently could be lowered to the ground from hover without the need for a great deal of cyclic or rudder pedal manipulating. The result was a very low workload throughout the approaches, landings and hovering. Sideward flight over the runway could be performed at speeds up to 35 kt. with no difficulty, as could left and right turns over a spot on the runway. Sikorsky test pilots said they have flown the S-76 in sideward flight at speeds up to 57 kt. right and 51 kt. left, but a 35 kt. limit was imposed because in some cases, the aircraft comes close to the 10% pedal margin required by the FAA.

Exceptional stability was again displayed in turns over a spot during the hover. The cross beam tail rotor, consisting of two crossed graphite composite paddles, offers greater strength than standard hinged, flapping tail rotor blades, Sikorsky engineers claim. This translates into greater tail rotor performance and better yaw control at high altitudes and/or in high crosswind situations. The simplicity of the rotor also eliminates lubrication requirements and makes the system almost maintenance-free, Sikorsky engineers claim. No bearings or elastomeric elements are used, and pitch changes are accomplished by twisting the paddles themselves with simple pushrods.

Landing from the hover also was surprisingly simple, no matter what direction the helicopter faced with respect to the wind. A slight downward push on the collective started the aircraft downward, without any perceptible need to make rudder or roll corrections on the way to touchdown.

Several features of the S-76 will make the aircraft as appealing to passengers as to pilots. These include:

- Head and legroom—In its high-density, 12-seat configuration, the S-76 still offers more leg room than most utility helicopters or high-density light fixed-wing transports now in service. The six-passenger executive version offers the equivalent of a small boardroom.
- Noise—While the utility version of the S-76 is not soundproofed to the extent that the executive version is, it was still possible to speak to passengers sitting diagonally across the cabin (a distance of more than 6 ft.) without having to raise one's voice significantly. Engine and transmission noise levels are much lower than on earlier helicopters, and executives accustomed to light twin-turboprop aircraft probably will not be able to tell the difference between the two, in terms of noise levels.
- Vibrations—New rotor technology and vibration damping techniques have contributed to a significant improvement in the ride of the S-76 over previous helicopters, even in such areas as translational lift on takeoff and landing. It is possible to write clearly and legibly throughout a flight in the S-76.
- Visibility—Visibility is good from all seats in the S-76, even from the center seats in the high-density version of the aircraft. The centerpost between the cockpit and cabin blocks forward visibility slightly from those seats, but sideward visibility is excellent.

From the pilot's point of view, it was difficult to find fault with the S-76. Seats were comfortable, cockpit configuration was convenient, visibility was very good and overall workload was low. Digital readouts for such performance parameters as engine gas generator speed and turbine exhaust temperatures made it easier to obtain maximum engine performance without overstressing the engines.

One feature that did not seem to work very well was the beep trim thumb switch on the top of the cyclic stick. Pressing the switch to trim the aircraft either longitudinally or laterally often appeared to have no effect on the attitude or flight path of the aircraft.

Part of this was due to the fact that the stick itself does not move when the trim system is used, and there is little sensation of any change in trim. In addition, the beep trim switch is limited to 3 sec. of operation. It must then be released and actuated again. Finally, the rate of trim change appeared to be too slow (a subjective issue that raises pilot debates from one new aircraft to the next).

From both the pilot's and the passengers' points of view, both the low vibration levels and the low noise levels are appealing. Much of the reduction in vibration is attributed to the aircraft's rotor system. The S-76 uses a forged aluminum main rotor hub with four barrels—one for each blade—that give it the appearance of a propeller hub. The aluminum hub was chosen because it offers simplicity in manufacturing and machining.

Titanium spindles on each blade are attached to the rotor hub through elastomeric elements in each barrel of the hub. The elastomeric elements absorb centrifugal, flapping and lead-lag loads from the blades, eliminating the need for sliding, oil-lubricated bearings.

One concession to this is separate lead-lag damper for each blade, which is lubricated from a common oil reservoir at the top of the rotor head. The dampers provide in-plane blade damping in ground conditions and in some flight conditions.

Above the main rotor hub is a bifilar rotor damper consisting of two four-arm plates with small weight at the end of each arm of the plates. The weights move in and out in opposition to blade forces to balance out vibration-inducing drag forces at higher speeds.

Fine-tuning of the aircraft's response to vibration is then accomplished through the self-tuning damper in the nose of the S-76. This damper senses changes in rotor rpm and adjusts itself to balance out any vertical motions in the nose of the aircraft. This keeps the avionics compartment smoother and quieter than it would be otherwise, Sikorsky engineers say, and results in greater reliability for the avionics.

Lightweight, corrosion-free composite materials have been employed extensively in the S-76. Program officials estimate that 40% of the surface area of the aircraft is made of composite materials, and most of the rear fuselage and tail is made of composites. Tail rotor itself is made of unidirectional graphite fibers, and weight is estimated to be about 60% of a conventional tail rotor.

Main rotor blades consist of several different types of composites. Blade tips are made of Kevlar. The blade spares are titanium, the interior structure is Nomex honeycomb and the skin is glass fiber. The blade roots are graphite.

Program officials concede that production costs may be somewhat higher because of this, but say there is an overall saving to the operator because of lower life-cycle costs. The blades, for example, are corrosion free with an unlimited life. Sikorsky is currently quoting a $1.1-million base price for the S-76.

S-76 Specifications

Maximum Speed	155 kt. (287 km./hr.)
Best Rate of Climb	1,350 ft./min. (6.9 meters/sec.)
Service Ceiling	15,000 ft. (4.573 meters)
Range/30 min. reserve	404 naut. mi. (748 km.)
Passengers	12
Crew	2
Engines	(2) Allison 250-C30
Weight Empty	5,273 lb. (2,392 kg.)
Gross Weight	10,000 lb. (4,536 kg.)
Useful Load	4,727 lb. (2,144 kg.)
Crew Seating Capacity (IFR Configuration)	2
Crew Seating Capacity (VFR Configuration)	1
Passenger Seating Capacity (IFR Configuration)	12
Passenger Seating Capacity (VFR Configuration)	13
Number of Blades—Main Rotor	4
Number of Blades—Tail Rotor	4
Passenger Cabin Volume	204 cu. ft. (5.78 cu. meter)
Baggage Compartment Volume	38 cu. ft. (1.08 cu. meter)
Passenger Cabin Area	45 sq. ft. (4.18 sq. meter)
Fuel Capacity (Std.)	280 U.S. Gal.

680 rotor system enhances Bell Model 222 performance

Nicholas C. Kernstock/Arlington, Tex.
May 15, 1989

The Bell Helicopter Textron Model 222 helicopter, equipped with Bell's 680 hingeless rotor system, displays the low vibration and high agility advantages of the rotor's glass fiber epoxy construction.

The 680 is the result of more than a decade of engineering development, according to Jim Harse, Bell's principal engineer on the project. A prototype of the rotor system was first flown on a Bell Model 222 technology demonstrator aircraft in May, 1982. Since then it has accumulated more than 800 flight hours and undergone several refinements such as addition of aerodynamically improved blades and a one-piece blade/cuff assembly (AW&ST Mar. 6, p. 61).

This *Aviation Week & Space Technology* pilot recently flew the 222/680 technology demonstrator aircraft at Bell's Flight Research Center at Arlington, Tex. Known as Plant No. 6, it is also the site for flight testing of the U.S. Marines' AH1W SuperCobra/680 rotor demonstrator aircraft (AW&ST Jan. 30, 1989, p. 19).

The basic aircraft is one of the first Model 222 prototypes built by Bell and was updated to include some of the modifications found on the 222B helicopter. Air fences on the bottom of the helicopter's sponsons were added to improve lateral stability, and the two Textron Lycoming LTS101 engine exhausts are extended and angled outboard to reduce exhaust gas reingestion problems.

The aircraft retains the original, shorter tail boom and small diameter tail rotor. A "Gurney flap"-type spoiler has been attached to the trailing edge of the vertical stabilizer to improve tail rotor off-loading in cruise flight.

Modifications specific to the experimental rotor system include:

- Removal of the "nodal-beam" transmission support and vibration isolation system.
- Installation of a solid transmission retention frame and two liquid vibration dampeners.
- Installation of a new, upper transmission case to accommodate the stronger, wider diameter mast.
- Elimination of the Bell stability control augmentation system (SCAS) pitch and roll channels. The yaw channel was retained.

In this photo sequence, the Bell Helicopter Textron 222/680 advanced rotor technology demonstrator rolls to the left over the Arlington, Tex., flight test area. The rotor system allows aerobatic maneuvers that could not be performed with the Model 222's original system.

Because the 222 cannot sustain inverted flight, it loses altitude rapidly as the aircraft passes through the inverted position. The resulting maneuver is similar to a barrel roll. Roll rates of up to 90 deg. per sec. have been achieved and the aircraft has performed full loops.

Although the yaw SCAS channel was still operational, it is not required for flight, and the aircraft was flown with a "bare" flight control system. This means that control feedback and vibrations are not automatically dampened out, and the pilot can evaluate more precisely the handling qualities of the rotor and airframe.

Most of the dynamic components on the aircraft are instrumented to record flight test data. A radio telemetry link can transmit real-time data to a ground monitoring station.

Dwayne Williams, a senior experimental test pilot at Bell, briefed me on the aircraft's systems and flight procedures. The program of maneuvers would demonstrate the rotor system's characteristics in both the military and civil flight regime.

Both the Marine SuperCobra and the Army's proposed LHX rotor systems will need a combination of high agility, quick response, low radar and acoustic signature, and tolerance to battle damage. Low-level attack and evasion operations and the new air-to-air mission require that attack and observation aircraft be capable of maneuvers not possible in previous helicopters. Aerobatic capabilities normally associated only with fixed-wing aircraft are being incorporated into designs for new helicopters.

After a brief walk-around inspection of the 222/680, I strapped into the left seat and Williams occupied the right seat. The cockpit interior was the same as the standard Model 222 with the exception of the g-meter and precision rotor tachometer, mounted above the right glare screen, and the mast torque indicator (MTI). The tachometer allows the pilot to accurately monitor and control the rotor rpm.

The MTI is mounted in the instrument panel and provides an indication of the torsional forces on the main rotor mast. Because of the rotor system's design, engine torque loads can be transmitted through the mast to the glass fiber blades and yoke with enough force to overstress the system, resulting in possible structural damage or reduced component life.

Bell engineers have installed a stick shaker on the collective pitch lever. The shaker activates at 92.5% transmission output torque and alerts the pilot that the MTI is approaching mast design limits when the pilot's attention is focused outside the aircraft.

After a normal engine start, Williams wheel taxied to the takeoff spot. Lifting the aircraft to a hover, he positioned us in a practice area. Williams gave me the controls, and I tried several pedal turns in both directions.

The difference between the 680 rotor and the standard two-blade rotor of the production 222 was immediately apparent. In addition to the almost total lack of rotor-induced vibration, the 680 is much more

Bell's Dwayne Williams, left, and Aviation Week & Space Technology *pilot Nicholas C. Kernstock prepare to start the two Lycoming engines.*

responsive and requires a delicate control touch. Hovering was made more difficult by a surface wind of 25–30 kt. After a few minutes of practicing, I was able to hold a stable hover.

Williams demonstrated a normal takeoff and climbed out. As the aircraft passed through about 30 kt., the shudder that usually signals the translational lift stage was almost imperceptible. In other rigid-rotor aircraft such as the Boelkow 105, this vibration is quite pronounced and, although short-lived, can be annoying.

We received approval for right traffic from the tower, and Williams retracted the landing gear. The downwind leg was flown at 100 kt. The aircraft was smooth and quiet throughout the remaining turns and approach to a hover. Once again, as the aircraft slowed through 30 kt. at about 20 ft. agl., the translational lift vibration was barely noticeable.

I took the controls and, after a clearing pedal turn, initiated a normal takeoff. At about 50 kt., Williams retracted the landing gear. The controls were crisp and stable in both the pitch and roll axis and felt similar to a fixed-wing aircraft.

The downwind leg was flown at 700 ft. agl. and 100 kt. Prior to turning to the base leg, I slowed to about 80 kt. and began a descent to 500 ft. Turning to the final approach at 500 ft. and 60 kt., I waited until we intercepted a 6-deg. approach angle and reduced the collective pitch firmly to initiate a descent. This usually would produce a definite increase in rotor noise and vibration with a conventional system. The 222/680 responded quickly with no noticeable vibration and only a slight increase in noise. The rest of the approach was stable and easily controllable.

After another traffic pattern, we requested a straight-out departure from the airport traffic area and flew to a test flight area about 10 mi. to the south of Plant No. 6. Williams took the controls and, when we had leveled off at 1,500 ft., slowed the aircraft to a complete stop to demonstrate vibration characteristics throughout the normal airspeed envelope. After holding an out-of-ground effect hover for a moment, he accelerated to 130 kt.

Feels solid in 60-deg. bank

As we neared the test area, I took the controls and began some basic maneuvers. The first was a deceleration to 50 kt. and then an acceleration back to cruise airspeed of 135 kt. while maintaining a constant altitude and heading. This requires the use of all flight controls and coordination of pitch and yaw attitude.

I then made a series of 360-deg. turns to the left and right, starting with a standard-rate, 30-deg. bank and then increasing the angle. In a 60-deg. bank, the aircraft felt solid with no tendency to flatten out of the turn or fall through, and I had no difficulty maintaining altitude and angle of bank.

I increased the pitch attitude sharply to initiate a cyclic climb. As the airspeed slowed to 60 kt., I executed a pushover. While still in negative g, I rolled left and right about 10 deg. The aircraft had very positive control throughout the pushover. We completed several more

All-composite rotor system consists of the black and orange blade-and-cuff assemblies and the cross-shaped white yoke. The gray cylinder above the yoke and small rectangular sensors are part of the test instrumentation package.

pushovers and pulled only .4g. This maneuver is not recommended in helicopters equipped with more conventional rotor systems.

Williams took the aircraft and demonstrated some of the aerobatic capabilities of the 222/680, beginning with a hammerhead. Starting at 120 kt., he pulled the nose up to the vertical and held that attitude until the airspeed fell to zero. As the aircraft began to slide backward, he applied right antitorque pedal, and the aircraft rotated around the yaw axis until it was pointing straight down. When the airspeed had increased to about 100 kt., Williams recovered with a 2.5g pull-out.

We climbed back to altitude and repeated the maneuver, this time executing a right vertical roll before the recovery. During the recoveries, the airspeed build-up was relatively slow and altitude loss was less than 700 ft.

Rolling inverted

I took the controls and climbed back to altitude to begin a series of split-Ss to each side. From straight and level flight at 90 kt., I rapidly applied hard left cyclic while momentarily pushing opposite antitorque pedal to maintain heading. As the aircraft passed through 90-deg. angle of bank, I neutralized the pedals and continued to roll inverted. Stopping the roll with neutral cyclic and left pedal, I gradually increased back pressure on the cyclic and monitored the g-meter to complete the half-loop recovery. Maximum force during the maneuver was about 2.9g. The aircraft is rated for as much as +3.80g and –.1g. At rest the g-meter reads 1.0.

My first split-S to the right produced a less than completely inverted position because the roll is more difficult in that direction. In the left roll, the aircraft rolls toward the blade on the retreating side.

One of the original Model 222 prototypes, the aircraft was modified to accommodate the 680 rotor system as well as several aerodynamic improvements. For example, the airflow fence, which is visible below the winglet, increases the aircraft's stability.

This blade normally has a higher angle of attack, while the advancing blade has less AOA. This is to compensate for the dissymmetry of lift produced as the rotor system moves forward through the air.

When the left roll is executed, the aircraft rolls in the direction of this asymmetrical lift and a higher roll rate is possible. The opposite occurs in a roll to the right. According to Williams, the 222/680 has a roll rate to the left of about 90 deg. per second.

Williams and I flew several more split-Ss and then executed vertical dives. After reducing the airspeed to zero kt. indicated, I pushed the cyclic forward until the nose of the aircraft was pointed directly at the ground. The airspeed built up to 100 kt. in about five seconds. We performed several of these maneuvers with half and full vertical rolls, initiating the recoveries at various airspeeds and g-loads.

We were unable to perform full loops and slow rolls because of a 2,000-ft. cloud layer over the test area. What is noteworthy is that the maneuvers we did execute required so little altitude to complete. By pulling 3.25g in the recovery, Williams executed a precise split-S with less than 500 ft. altitude loss. Less extreme aerobatic maneuvers can be accomplished in as little as 100 ft.

We then descended to tree top altitude to evaluate the aircraft's nap-of-the-earth performance. Current military doctrine calls for helicopters to operate regularly at altitudes of 4 meters and below. This requires not only lateral maneuverability but also a high degree of pitch agility.

The 222/680's controllability in pushovers and high roll rate allow it to closely follow terrain contours. Another advantage of the composite rotor is its ability to dampen out turbulence. The surface winds in the test area were gusting to 35 kt., but the mechanical turbulence was nearly unnoticeable inside the aircraft.

We flew back to Plant No. 6 at 1,500 ft. and entered right downwind traffic. As we rolled out on final, Williams lowered the landing gear and I landed the aircraft.

The rotor system exhibited fine flying qualities and low vibration and noise levels throughout the flight. While its stability would make it an effective IFR platform, its agility and responsiveness make it particularly attractive on a combat aircraft. The lack of vibration also could improve the reliability of avionics on both civil and military aircraft.

The 680 rotor has been incorporated in the McDonnell Douglas/Bell Helicopters entry in the Army's LHX competition. It also could have wide applicability to other new designs as well as upgrades of existing helicopters.

Engineers at Bell believe the 680 rotor could have wide potential as an upgrade for existing civil helicopters, although the company has no immediate plans to market the system. In addition to the Bell Model 222 pictured here, it is undergoing flight tests on a Marine Corps AH-1W.

Bell presses Model 222 certification

Robert R. Ropelewski/Los Angeles
April 2, 1979

Federal Aviation Administration is expected to issue a type inspection authority for the Bell Helicopter Textron Model 222 corporate/utility helicopter this week, opening the door for final certification flight testing of the 222 by FAA test pilots and an anticipated type certification of the aircraft in mid-May.

Bell officials earlier had expected to obtain the type inspection authority for the Model 222 in late January, but a series of delays resulted in continued postponement of the authorization. A Bell executive said the delays could not be attributed to any single item in the program, but rather were due to a number of issues that reflected the company's conservative approach to the certification process.

Bell has made significant advances in the areas of stability, comfort and safety with the Model 222, while continuing to rely on the relative simplicity of a two-bladed rotor system.

This *Aviation Week & Space Technology* editor flew the Bell 222 recently, and found it exhibited—like other new-generation helicopters entering the civil marketplace—near fixed-wing standards of noise, vibration levels, stability and overall comfort. The two-segment evaluation flight included time in both the cockpit and the passenger cabin, allowing observation of the 222 from both the pilot's and the passenger's points of view.

Bell 222 corporate and utility helicopter is scheduled for certification this month and deliveries of the aircraft are due to start in October.

Bell Helicopter Textron Model 222s are assembled at the company's facility in Ft. Worth, Texas. The Model 222 is powered by two Lycoming LTS101-650C-2 turboshaft engines.

Aircraft used for the evaluation was the third of five 222 prototypes now flying, which together have logged nearly 2,000 flight test hours since the first prototype flew in August, 1976. Bell engineers anticipate these aircraft will log another 1,000 hr. in flight testing before production deliveries begin in October.

Stability of the 222 is such that Bell expects to certificate the aircraft for dual-pilot instrument flight rule (IFR) operations with only a magnetic brake force trim system to meet FAA requirements for a positive stick force gradient on the cycle control stick. Dual stability augmentation systems (SAS) will be offered as an option, however.

The No. 3 aircraft flown for the *Aviation Week & Space Technology* evaluation was fully equipped with dual flight directors, dual SAS, weather radar and a heading hold feature in the SAS yaw channel that had been installed for Coast Guard evaluation in conjunction with that service's SRR (short-range recovery) program.

Externally, the main features of the 222 noted during a preflight walk-around included:

- Two-bladed main rotor system with a hub that Bell engineers say is the simplest ever designed by the company. No stabilizer bar is used, as it is on most of Bell's earlier two-bladed systems, and the overall number of parts is lower than on any previous Bell rotor system, despite greater redundancy in the 222 system.

- Sponson-type housing for fuel tanks and landing gear, located just behind the passenger cabin in either side of the fuselage. Optional extensions will be available to house additional fuel and emergency popout flotation systems.
- Retractable tricycle landing gear with a nosewheel that retracts forward into the nose, and main gear that retract into the sponsons on either side of the fuselage.
- Fixed horizontal stabilizer with slotted leading edge and large vertical stabilizer that has two smaller auxiliary fins or the horizontal stabilizer to unload the tail rotor at high speeds.
- Large, one-piece nose-cone that tilts down and forward to provide access to an equipment compartment with 14 cu. ft. of space for avionics, electrical equipment and a radar antenna.

The fuel tank/landing gear sponsons on each side of the aircraft are located near the engines and rotor mast of the 222, and thus provide work platforms for inspecting and maintaining most of the dynamic components of the aircraft.

Bell has chosen short, wide rotor blades that can be operated at tip speeds about 14% lower than previous two-bladed systems, resulting in reduced rotor noise. Tail rotor speeds have been similarly slowed to reduce tip noise.

Composite materials are employed extensively in the Bell 222 rotor system and blades. Primary load carrier in the blades is a stainless steel spar, but redundancy is provided, Bell engineers say, by bonding glass fiber safety straps along the upper and lower inner surfaces of the spar to retard the propagation of any cracks and offer a secondary load path for the blades. The straps wrap around each blade attachment bushing and arc capable of carrying the blade centrifugal loads.

Noncorroding Nomex honeycomb core covered with a glass fiber skin is used for the afterbody of the blades, and a replaceable stainless steel leading edge strip protects the blade spar at the front of the blade.

Main rotor hub carries centrifugal and bending loads through a titanium yoke, with unlubricated elastomeric bearings absorbing the centrifugal loads, blade feathering motion and teetering or flapping motion. Torsion springs on the hub limit teetering at low rotor speeds, thus allowing increased ground and tailboom clearances during the startup and shutdown in gusty wind conditions.

Behind the rotor mast, upward-opening cowlings permit access to the two Lycoming LTS101-650C-2 turboshaft engines powering the 222, and engine inspection and maintenance can be performed while standing on the sponsons.

Interior of the No. 3 prototype is in a seven-place configuration, including two crewmembers' seats in the cockpit. The passenger compartment accommodates two rear-facing seats just behind the crew seats, and three forward-facing seats at the rear of the compartment.

Standard 222 configuration offers a deluxe eight-place interior with two seats in the cockpit and six forward facing seats in the aft cabin. Other interior arrangements are possible, including a high-density 10-place configuration, a cargo configuration with all seats but the pilot's removed, and an ambulance configuration with litters. There is a 23-×-55-in. opening between the passenger cabin and the baggage compartment that provides a straight-through space for carrying items up to 12.5 ft. long.

Cockpit configuration of the No. 3 prototype was much like that of a fixed-wing business aircraft, with dual controls, a full-width instrument panel, and dual IFR instrumentation. This particular aircraft had a non-standard avionics package, but production aircraft will be offered with several avionics options built around the Collins Pro Line and Micro Line series of navigation and communications radios.

Avionics installation

Space is reserved for Pro Line avionics control heads at the top center of the instrument panel, while Micro Line equipment, if used, still will be located at the left center of the panel, next to engine and transmission gauges.

Space is reserved for a weather radar at the forward end of the center console, and Bell officials believe that at least 35–40% of all 222s sold will be equipped with radar.

Probably the most unique feature of the 222 cockpit is the fore-and-aft collective located on the right (pilot's) side of the cockpit. Standard floor-mounted collective with dual twist-grip throttle handles at the end is used on the left side.

The fore-and-aft collective for the pilot in the right seat is built into the right side of the center console. Rather than a long stick, the collective is more of a short, upright pole with the twist-grip throttles for the two Lycoming LTS101 engines protruding from the upper left side of the pole.

Floor space on the left side of the right seat is clear, allowing easier movement between the seat and the passenger cabin. The side-by-side throttles are easier to associate with their respective engines than those on a standard collective, and their position over the center console puts them directly beneath the gauges for their respective engines on the front instrument panel.

Bell test pilots claim one of the biggest benefits of the new collective configuration is that it eliminates the vertical-mode pilot-induced-oscillations that often occur when collective friction is too loose. Problems were encountered with this on the Bell 212 helicopter. The new collective design will be incorporated in Bell's Model 214ST, now in development.

Cockpit seats in this aircraft had a limited range of travel, both up and down and fore-and-aft. No matter how far down and aft the seats are put, it is difficult for an average or tall person to read the top row of instruments on the instrument panel. Roger Huffaker, the Bell 222 test pilot with whom the *Aviation Week* evaluation flight was flown, noted that seat adjustment limits will be extended on production aircraft, and the instrument panel will be tilted upward slightly for better visibility of the numbers at the top of instruments and gauges.

Starting and rotor turn-up was accomplished with onboard battery by first arming the "start" switch for each engine on the overhead panel, then momentarily pushing a "start" button on a small console on the pilot's collective. Rotor brake is not used while the engines are running, and the rotor began turning as the first engine was started.

As engine gas generator (N_1) climbed through 10–12%, the throttle for that particular engine was rolled off the "stop" position toward the "flight idle" position. Between the two positions, the throttle was adjusted to keep turbine inlet temperatures from exceeding their limits.

Once both engines were running, rotor rpm was increased to 100% using a beep trim switch on the small console on the pilot's collective. Rotor speed on this aircraft at 100% was about 338 rpm, but this will be increased to 348 rpm on production aircraft to improve hover performance, allow higher airspeeds and provide better tail rotor control.

Taxiing was initiated by releasing the parking brakes and tilting the cyclic stick forward slightly. There was no need to increase collective pitch. This was helpful because the 222 was parked on a flight line with other aircraft nearby.

Rudder pedals on both sides of the cockpit were equipped with toe brakes, and the brakes were effective without being overly sensitive. Maneuverability on the ground was excellent, using the rudder pedals to control tail rotor torque in turns. The 360-deg. castoring capability of the nosewheel permits extremely tight turns.

Aircraft weight

Aircraft weight just before liftoff was about 7,300 lb., including around 400 lb, for the two pilots and 1,100 lb. of fuel. Empty weight of the No. 3 prototype is 5,795 lb., but empty weight of production Model 222s in standard configuration is expected to be around 4,550 lb. The aircraft will be certified to a normal gross weight of 7,650 lb. and an external gross weight of 8,200 lb.

With an airport altitude of about 800 ft., temperature around 59F and a wind of approximately 12–15 mph, the 222 was lifted into a hover. Stability was very good both with stability augmentation on and off, although the workload required to maintain a stable hover was noticeably higher with SAS off.

Bell uses a mast torque meter for power reference in the Model 222. This is a reading of the torque applied to the rotor mast at the transmission, rather than a reading of the engine torque being applied to the transmission. Usefulness of this is that it gives a direct indication of the power going to the main rotor without having to consider that portion of available engine power going to the tail rotor. In the 222, the same gauge that displays mast torque also shows torque from each of the two engines at the transmission.

In this case, only about 70% of mast torque was needed to maintain our hover.

With stability augmentation systems or additional power applied, the nose was lowered slightly and the aircraft accelerated quickly to a climb speed of 65–70 kt. Vibrations were negligible in passing through translational lift during the acceleration. Landing gear was raised and the aircraft stabilized in a climb of about 1,900 ft./min. at 100% mast torque.

Leveling off at 2,000 ft., mast torque was set at 87%, which was within the 222's economy cruise power range of 85–90%. Airspeed stabilized at around 130 kt. at 90% mast torque. Cruise speed would have been about 145 kt., according to Bell's performance charts for the 222. Maximum allowable speed for the 222 is 160 kt.

Hands-off flight

Heading and altitude hold were engaged briefly on the autopilot, and a hands-off flight capability was demonstrated. Shortly thereafter, the autopilot was disengaged and we climbed to 3,000 ft. to overfly some mountains.

Despite the moderate winds blowing in the area and the resultant air turbulence coming off the mountains, there were no noticeable rotor excursions and the aircraft continued to fly smoothly. Bell uses a

Nodamatic suspension system for the rotor and transmission in the 222, consisting of a pair of composite-structure beams plus tuning arms and weights.

The transmission is suspended from the two beams, which then absorbs vibrations from the rotor. The vertical vibrations from the rotor act over most of the length of the beams, except for a point near each end of the beams. These "nodal points" can be fine tuned with the weights to minimize vibrations, and it is from these points that the fuselage is attached to the beams.

The result as seen on this flight was a smooth ride up to about 140 kt. There was a definite rise in the amplitude of two-per-rev vibrations at speeds above 140 kt. These increased steadily up to 160 kt., inducing a noticeable shaking of the instrument panel. Huffaker and Bell engineers said much of this will be eliminated on production aircraft, which will have a door on the left side of the cockpit. This will take some of the stiffness out of the forward fuselage area.

Stability remained exceptionally good at all speeds, with SAS both on and off. Extended periods of hands-off flight were possible either way, although as airspeed increased, rudder corrections were needed more frequently with SAS off.

Cyclic and rudder trimming was accomplished through the force-trim release trigger on the forward side of the cyclic grip. Control forces were slightly higher, though still very comfortable, with the force trim engaged, and aircraft response to control inputs was noticeably quicker and more precise than with the force trim disengaged.

Turning off the No. 2 hydraulic system brought no distinguishable change in the aircraft's flying qualities.

The main disadvantage in the loss of that particular system comes from the fact that it powers the landing gear, and the landing gear must be lowered manually if the system is lost. Landing gear itself can be lowered raised or left extended up to 140 kt. There were no trim changes needed when the gear was lowered or raised.

While cruising at 2,500 ft., Huffaker shut off the No. 1 engine. The aircraft yawed to the left slightly with the loss of torque, but that was corrected easily with the rudder.

Using a beep switch located on the pilot's cyclic console, gas generator speed on the remaining engine was increased to its maximum of 104.8%. With 87% torque from the No. 2 engine and 60% mast torque, the aircraft maintained its altitude and stabilized at an airspeed of around 110 kt. Performance charts for the 222 indicated that under prevailing weight ad temperature conditions, it should have been possible to climb to about 10,200 ft. although the single-engine service ceiling in standard conditions and at a gross weight of 7,000 lb. is listed at around 9,400 ft.

The No. 1 engine was restarted, then both engines were brought to idle to simulate a duel engine failure. Airspeed was slowed to 70 kt., which is the best rated descent airspeed for the 222, and the aircraft entered a comfortable autorotation that stabilized at a descent rate about 1,700 ft./min.

Residential areas below necessitated a recovery from the autorotation several hundred feet above the ground, and we climbed back up to 2,000 ft. to return to Van Nuys Airport. Final approach speed was flown at 60 kt., and as the aircraft slowed from that speed to a hover, there was a slight shuddering as it passed through translational lift again.

On the ground, two additional passengers were picked up for a demonstration ride and we took off again. This time, this *Aviation Week* editor sat in the passenger cabin while Huffaker flew the aircraft.

Weight of the two additional passengers compensated for the fuel burned, and overall weight of the aircraft was approximately the same as it was for the first segment of the flight. Cruise was established at 2,500 ft., with 85% mast torque providing 140 kt. airspeed.

Short climb was initiated to 3,000 ft., where the aircraft was hovered with a 75% mass torque reading. Performance charts for the 222 indicate it could have hovered out of ground effect up to an altitude of approximately 8,000 ft. At our estimated weight of 7,300 lb. in ground effect, the hover ceiling at that weight and temperature should have been around 10,800 ft.

It was quiet, comfortable and smooth in the cabin, and easy to talk to the two other passengers without shouting. Bell engineers say a multifaceted approach was taken to reducing cabin noise, including the short, wide blades with low rotor and tail rotor tip speeds, selection of equipment that offered the lowest noise levels, lower speeds for hydraulic pumps to reduce their characteristic whine and the use of elastomeric bearings to mount the transmission to the fuselage through the Nodamatic beams.

Engines shut down

At the end of the flight, engines were shut down and the rotor brake was used to stop the rotor. The rotor can be stopped in winds up to 55 kt.

Bell engineers have incorporated several safety features in the 222, including:

- Shoulder harnesses on every seat.
- Crash-resistant fuel cells. These have been demonstrated not to rupture when filled with water and dropped from a height of 50 ft.
- Frangible, self-sealing connections between the three fuel tanks in the fuselage and the fuel tanks in each stub wing.

- Suction fuel system, rather than a boosted-feed or gravity system, so that in case of an accident, fuel pumping will stop with engine shutdown or with disengagement of fuel lines to the engines.
- Energy attenuating seats with cylindrical honeycomb aluminum rolls underneath to help decelerate loads smoothly. In tests of this concept, crash loads of up to 20g were imposed on a 170-lb. dummy in one of these seats. Those conducting the tests determined that the seats absorbed the entire 20g, and a person sitting in the seat in a real crash of that magnitude would not have incurred back injury if he or she did not already have a history of back problems.

The first production 222 is now on the flight line at Bell's Ft. Worth plant, and is expected to start flight tests at the beginning of June. That aircraft will be used to flight test those aspects of the aircraft that have been modified from the prototypes.

Bell Model 222 Performance Summary

(Standard Day, Except as Noted)

Takeoff gross weight (lb.)	6,800	7,200	7,650
IGE. hovering ceiling ft.	14,000	12,200	10,300
OGE. hovering ceiling ft.	10,400	8,500	6,400
IGE. hovering ceiling, hot day* (ft.)	10,700	8,300	6,700
OGE. hovering ceiling, hot day*(ft.)	6,700	4,500	2,100
Certificated altitude (ft.)	TO BE DETERMINED		
Service ceiling max. climb pwr. (ft.)	17,900	16,200	14,300
Maximum rate of climb, sea level, T/O pwr. (ft./min.)	2,160	1,950	1,730
Vertical rate of climb, sea level, T/O pwr. (ft./min.)	1,740	1,330	860
Max. cont. cruise, TAS, sea level (mph.)	170	168	165
Economical cruise, sea level (mph.)	152	151	150
Range, sea level (mi.)	411	404	395
Economical cruise at 4,000 ft. (mph.)	152	151	150
Range at 4,000 ft. (mi.)	453	442	425
Economical cruise at 8,000 ft. (mph.)	150	148	146
Range at 8,000 ft. (mi.)	489	470	450
One engine inoperative, service ceiling, 30 min. power (ft.)	10,300	8,500	6,400
One engine inoperative, service ceiling, hot day*, 30 min. power (ft.)	6,800	4,700	2,300

*Hot day is standard day +20C.

BK.117 displays engine-out margins

Robert R. Ropelewski/Los Angeles
June 4, 1984

Responsive flight controls and comfortable performance margins with one engine inoperative are among the key attributes of the Messerschmitt-Boelkow-Blohm/Kawasaki BK.117 twin-engine, 8-11-place helicopter. Additional improvements under consideration for the aircraft will further strengthen its performance and competitiveness in the light twin-engine helicopter field (AW&ST Mar. 19, p. 103). This *Aviation Week & Space Technology* pilot flew the BK.117 on a 1-hr. 20-min. demonstration flight and found it to be a smooth-flying, relatively high-performance aircraft.

Design concepts

The aircraft's handling characteristics and flying qualities are similar to those of MBB's smaller BO.105 helicopter, from which many of the BK.117's design concepts were taken.

The flight was conducted from Van Nuys Airport near Los Angeles, with Robert Cuevas of MBB, Inc., MBB's U.S. subsidiary, serving as the instructor and familiarization pilot. The aircraft had logged 115 hr. prior to the start of our flight.

Cockpit of the BK.117 is as efficiently arranged as the BO.105's. Although the newer aircraft is not yet certificated for instrument flight rule (IFR) operations, the demonstrator aircraft used on the *Aviation*

MBB/Kawasaki BK.117 helicopter was flown on Aviation Week & Space Technology *evaluation flight during West Coast demonstration tour.*

Week flight was fully equipped with dual vertical gyro indicators, horizontal situation indicators and VHF omnirange receivers as well as gin Arnav 60 Loran-C low-frequency radionavigation receiver. I took the right seat of the aircraft for the flight, while Bob Cuevas occupied the left seat.

The BK.117 is powered by two Lycoming LTS101-650B-1 turboshaft engines rated at 650 shaft horsepower (shp.) each. The engine controls are located on the center overhead panel in an arrangement similar to the BO.105. I started the right engine first by pressing the start button, then moving the engine power lever forward slightly when compressor speed (N_1) reached 10% to introduce fuel into the engine. The lever could then be moved fore or aft as necessary to maintain interturbine temperature between 600C and 700C while the engine spooled up. At 40% N_1, I released the start button and pushed the power lever into the "idle" position.

Rotor turn-up during the engine starts was relatively smooth, and there was no significant shaking or rocking of the aircraft during the turn-up. With the rotor at flight rpm and the rotor blades at flat pitch, the engines were operating at about 81% N_1.

Weight of the aircraft at that point was approximately 5,600 lb., which included an empty weight of 4,250 lb., 990 lb. (450 kg.) of fuel, and our combined crew weight of about 360 lb. The normal maximum gross weight of the BK.117 is 6,283 lb.

Conditions were close to international standard atmosphere (ISA), with a temperature of 60F, barometric pressure of 30.10 in. Hg. and negligible wind. Van Nuys Airport is at an elevation of 799 ft., and the aircraft lifted easily into a hover under those conditions using about 40% torque on each engine.

Maximum power allowable per engine is 71% torque with both engines operating. With one engine inoperative, the remaining engine can be operated for up to 12 sec. at 100–120% torque, up to 2.5 min. at 100% torque, up to 30 min. at 91.5% torque and continuously at 83% torque. Under prevailing conditions, loss of one engine in the hover would have resulted in the remaining engine going to 80% torque, still within the maximum continuous operating range.

Smooth acceleration

Cuevas flew the aircraft on the departure from Van Nuys, using about 60% torque on each engine for the takeoff. The aircraft accelerated quickly and smoothly as we left the Van Nuys terminal area.

We leveled off at 2,300 ft. without changing the power setting, and the aircraft stabilized at an indicated airspeed of 139 kt. After taking the controls, I raised the collective level to increase the power to the maximum allowable of 71% torque on each engine. Our indicated airspeed increased to 145 kt., close to the BK.117's sea level true airspeed limit of 150 kt. The vibration level remained low at that speed, although noise from the airflow around the fuselage increased noticeably.

A four-direction trim switch on the top of the cyclic stick allowed the aircraft to be trimmed in any desired attitude, thus serving as a basic short-term attitude hold system. The BK.117 has no stability augmentation system and, for visual flight operations, does not appear to need one.

Cuevas pulled the left engine power lever to the ground idle position to simulate an engine failure. The aircraft yawed slightly to the right, as the right engine absorbed the additional power requirement, but this was easily corrected and posed no significant control problem.

Engine out

Cuevas suggested slowing the aircraft to 100 kt. or less with one engine inoperative because the nose has a tendency to tuck downward

MBB demonstration pilot Robert Cuevas inspects avionics compartment of BK.117. Most key components are easily accessible from the ground or from built-in workstands on the aircraft.

slightly if the remaining engine is lost and the collective pitch lever is lowered suddenly at higher speeds than that. In addition, a 90-kt. steady-state autorotation speed is recommended.

I continued to slow the aircraft to 65 kt., the best single-engine climb speed, then raised the collective pitch lever to increase power on the right engine to the maximum allowable of 83% torque. The aircraft stabilized in a 750-fps. climb with the interturbine temperature on the right engine at about 740C, comfortably below the 2.5-min. limit of 832F.

Cuevas took the controls briefly as we reached 3,000 ft. and brought the BK.117 to a single-engine hover at that altitude. With no wind, the aircraft showed a tendency to settle slightly at the maximum continuous single-engine power limit of 83% torque. The settling was stopped with a power setting on the right engine of about 86% torque, which has a 30-min. limit, before Cuevas pushed the nose down slightly and we accelerated out of the hover. With both engines operating, the BK.117 at maximum gross weight on a standard day can hover out of ground effect up to 8,200 ft., according to performance charts for the aircraft.

Sensitive controls

Taking the controls again after Cuevas reestablished the aircraft at cruise speed with both engines operating, I executed a series of shallow and moderate-bank turns to continue getting a feel for the aircraft. The controls of the BK.117 are sensitive and responsive, similar to those of the BO.105, which I had flown on several previous occasions, and in contrast to many other helicopters. While this makes the BK.117 a more enjoyable and precise machine to fly, it takes a while to become accustomed to the control sensitivity. Some of my initial hovering work in the aircraft was not completely stable until I began making much smaller control movements to counterdrift in all three axes.

During approaches to a hover, the BK.117 has the feel of a much heavier aircraft, despite its control sensitivity, and gives the sensation of being slower than it actually is on landing approaches. I found that frequent glances at the airspeed indicator were necessary to sustain a steady airspeed reduction and avoid an exaggerated flare over the landing zone.

Another reason for keeping a steady airspeed reduction rate on the approach was to pass quickly through the regime from approximately 30 kt. down to 25 kt., where a distinct airframe shuddering occurs as the aircraft passes through translational lift. These vibrations

Cockpit of the BK.117 demonstrator aircraft is well equipped for both visual and instrument flight rules operations, although the helicopter has not yet been certified for IFR service. Certification for both dual pilot and single pilot IFR operations is expected this year.

were more noticeable in the approaches because the BK.117 was relatively smooth throughout the rest of the flight envelope, including passage through translational lift on take-off and in high-speed cruise.

The airframe remained fairly level and the nose relatively low in the final flare to bring the aircraft to a hover. These characteristics remained constant in several approaches at various speeds and descent angles to a 2,800-ft. mountain top. About 45% torque was needed from each engine to hover in ground effect at that altitude.

The absence of any significant winds facilitated directional control in the hovers. At an altitude of approximately 6,600 ft. and above, the BK.117 has a critical wind component of 17 kt. from 90 deg. right due to large vertical end plates on the horizontal stabilizer. The end plates help unload the tail rotor at cruise speeds, but at low speeds and in the hover they complicate directional control because of the large area exposed to crosswinds. MBB is testing a wider-chord tail rotor blade that will provide more directional control, and it also is planning to reset the vertical end plates on the horizontal tail from a present 8 deg. angle of incidence to a 5 deg.

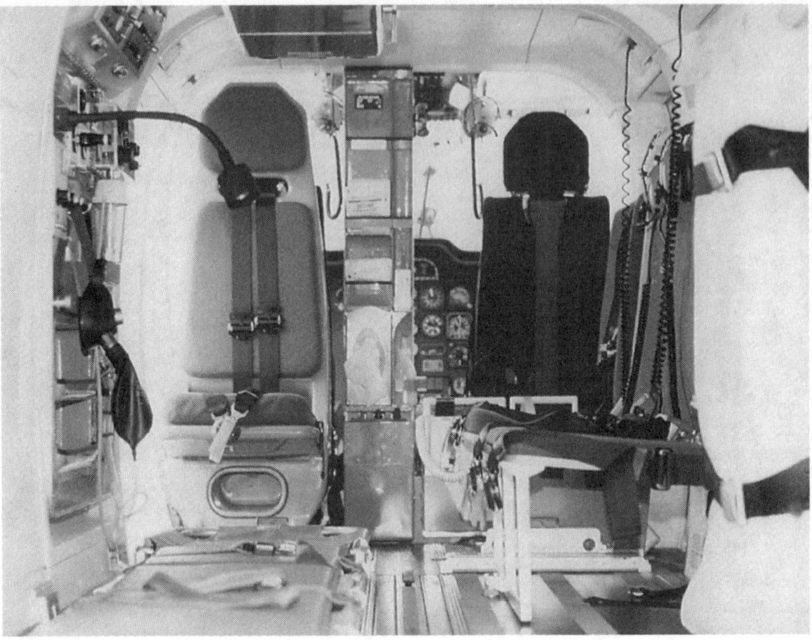

BK.117 flown on Aviation Week *evaluation flight was equipped in emergency medical services configuration, with lightweight Heinrich Bucher EMS system developed originally for the Swiss Air Rescue Service. Two seats for medical attendants are installed at forward end of the cabin. Stretcher and emergency medical equipment are at left; another stretcher is at right.*

Holding a steady heading in a hover was not especially difficult during our flight because the winds were negligible. One complication that did arise was a slight binding of the rudder pedals during hovering turns over a spot on the ground. This occurred when initiating or stopping the turns, when the left rudder pedal was about 1 in. farther forward than the right pedal. At that point, a stiffening in the controls occurred, which necessitated a harder push on the right rudder pedal to keep it moving forward. This resulted in some mildly erratic nose movements when the resistance ceased and the right pedal jumped forward. This condition was later found to have been caused by a faulty bushing in the tail rotor control system. The bushing subsequently was changed, eliminating the problem.

Sideward flight characteristics of the BK.117 were conventional. The aircraft is limited to a speed of 20 kt. in sideward and rearward flight. Takeoff accelerations from the hover were fast and smooth, taking about 20 sec. to reach 100 kt. Engine response was excellent throughout the flight, and at no time did the LTS101 engines allow rotor speed to vary more than a percent or two from our originally selected rpm, no matter how quickly power settings were changed.

Cuevas made the final landing at Van Nuys, keeping our airspeed at about 60 kt. on final approach, then slowing rapidly to pass quickly through the 30–25-kt. vibration zone. We landed 1 hr. 20 min. after our initial takeoff, with about 330 lb./150 kg. of fuel remaining, for a total fuel burn of about 660 lb./300 kg.

After allowing the engines a 2-min. cool-down period at idle power, Cuevas brought the overhead engine power levers to the shutoff position. A rotor brake handle was located on the floor to the right of my seat, and I applied the brake as rotor rpm dropped below 50%. At Cuevas' suggestion, I released the brake momentarily just before the rotor came to a complete stop to avoid the shock of a sudden stop on the rotor system.

Hingeless rotor

Both externally and internally, the BK.117 offers many similarities to the smaller BO.105. The hingeless rotor system of the newer aircraft is essentially that of the five-passenger BO.105, although rotor blades have been lengthened by 2 ft. to give the BK.117 a 36.1-ft. rotor diameter compared with 32.1 ft. for the BO.105. The BK.117 blades also have a slightly wider chord than those of the BO.105. The fuselage and tail boom of the BK.117 closely resemble those of the BO.105, although the tapered nose helps give the aircraft a more streamlined appearance than the BO.105 with its rounded nose.

Rear clamshell doors on the BK.117, also adopted from the BO.105, facilitate loading and unloading of the cabin when the aircraft is used for tasks such as cargo transport or medical evacuation. In the emergency medical services configuration, for example, one or two stretchers can easily be slid in and out of the cabin from the rear.

The EMS configuration installed in the BK.117 demonstration aircraft was developed and manufactured by Heinrich Bucher, a Swiss firm, and is the same kit used by the Swiss Air Rescue Service. The configuration included dual stretchers, emergency life support equipment, and rear and side-facing seats for two medical attendants just behind the pilot's and copilot's seats.

A relatively flat, two-stage transmission designed especially for the BK.117 fits on the roof of the aircraft, thus giving the cabin interior a boxcar effect with no obstructions. This gives the cabin about 200 cu. ft. of uninterrupted space.

MBB and Kawasaki have had only limited success so far in selling the BK.117 in the U.S. and elsewhere because of general economic conditions and the downturn in demand for business and general aviation aircraft in recent years. One area in which the German/Japanese aircraft have generated some relatively strong interest is the emergency medical services field, and it was an EMS-configured version of the BK.117 that was used on the *Aviation Week* flight. The aircraft flown was the 11th BK.117 to be produced.

Because of the relatively compact exterior dimensions and spacious and unobstructed interior of the BK.117, MBB and Kawasaki are marketing the aircraft for a variety of applications, including corporate/executive transportation, offshore oil drilling support, on-shore seismic exploration, and emergency medical services. EMS has been a small but active market over the past two years, and the BK.117 has drawn strong interest.

Sales turnaround

As with most other helicopter manufacturers, MBB and Kawasaki are awaiting a turnaround in other markets to bring sales back to late 1970s/early 1980s levels. The two companies delivered 20 BK.117s worldwide in 1983, and anticipate bettering that number this year. In the meantime, the two firms are working on further improvements to the BK.117, including an increase in gross weight from 6,283 lb. to 7,040 lb., a more powerful tail rotor and higher power engines that would increase high/hot performance of the aircraft.

Hughes 500D demonstrates versatility

Donald E. Fink/Culver City, Calif.
February 14, 1977

Hughes 500D, the latest commercial design Hughes Helicopters has developed from its U.S. Army OH-6 light observation helicopter, retains the agility of earlier models while offering significant performance improvements with its higher thrust engine, upgraded power train and fifth main rotor blade.

Maximum gross weight of the 500D, which is powered by the 420-shp. General Motors Allison 250-C20B engine, is 3,000 lb., compared to 2,550 lb. for the C model. The earlier version has a four-bladed main rotor and is powered by the 400-shp. 250-C20 engine. The increased installed power in the 500D, the fifth rotor blade and a new T-tail for improved high-speed longitudinal stability, give it a maximum never exceed (V_{ne}) speed of 156 kias.

At its maximum gross weight, the D model has a top cruise speed of 140 kias., a hover ceiling out of ground effect of 7,400 ft., a service ceiling of 15,200 ft. and a sea level endurance of 2.8 hr. The rate of climb at sea level and takeoff power is 1,680 fpm, and the vertical rate of climb under the same conditions is 910 fpm.

Other external changes made on the D model include a 4-in. increase in the diameter of the tail rotor and a 2-in. extension of the tail rotor boom. Both changes were needed in order to compensate for the increased power delivered by the 250-C20B engine.

An optional four-bladed tail rotor, with the blades canted instead of perpendicular for improved noise characteristics, has been developed for the 500D (AW&ST Jan. 20, 1975, p. 40). The new tail rotor will be certified later this year, and will be offered both as a production option and for retrofit on earlier production models.

Power train changes made on the D model include an uprated main transmission that has new gear ratios, stronger bearings and an improved lubricating system. A heat exchanger also has been added to cool the transmission oil. On the C model, transmission oil cooling is achieved by circulating ambient air around the transmission.

Takeoff power in the 500D is limited by the transmission to 375 shp.—a power setting that can be sustained for 5 min. Maximum turbine outlet temperature allowed for takeoff power is 810C. Maximum continuous power rating is 350 shp., with a turbine outlet temperature limit of 738C.

Hughes is offering the 500D as an extension of its line of versatile lightweight helicopters, which are suitable for business transport and yet retain the ability to perform a variety of utility tasks—including maximum gross weight operations at an altitude of 4,000 ft. on a 95F day. It can lift an external sling load of 2,115 lb., when operated in a special utility configuration at a maximum gross weight of 3,700 lb.

Military versions of the helicopter, embodying the same basic design improvements, also are under development for the export as well as domestic market. A total of 250 D models have been ordered—150 commercial versions and 100 armed 500MD versions for South Korea.

Performance of the 500D was evaluated here recently by this *Aviation Week & Space Technology* editor, flying two demonstrator models on different days and under different conditions. This pilot earlier had checked out in the smaller Hughes 300C, which is powered by the Avco Lycoming HI0360-D1A reciprocating engine. The experience in the 300C provided valuable exposure to the Hughes design philosophy, which stresses use of a lightweight airframe with a low-mass, low-inertia rotor system.

The first 500D flight was made with this editor flying in the left command pilot position and Walt Hodgson, a Hughes experimental test pilot and instructor pilot, in the right seat. Following this 1-hr. familiarization flight, a second 500D was flown for 1 hr. 6 min. with Chuck Hench, also a Hughes experimental test pilot, flying in the left seat.

Objective of the first flight was an introduction to the 500D, which this pilot had not flown before, and an initial checkout in basic flight maneuvers. The helicopter was operated at the reduced 2,200-lb. gross weight, which made it possible for Hodgson to demonstrate the machine's agility with a series of low-altitude, high g-force maneuvers.

The second flight was aimed at demonstrating the 500D's guaranteed performance, and 800 lb. of lead shot ballast was strapped onto the floor of the aft cabin to achieve the 3,000-lb. maximum gross weight. Hench flew precise engineering points, with this editor observing and recording the points achieved.

Preflight walkaround inspection for the 500D does not differ much from that used for earlier models, and is similar to that followed on most helicopters. Checking the oil levels in the main and tail rotor transmissions is facilitated by indicators that are readily visible from ground level.

The aft lower cone of the fuselage consists of hinged panels that open to provide access to the engine compartment for that portion of the preflight inspection. The interlocking panels are held closed by spring latches along the bottom fuselage line.

The rotor head inspection includes checking the dome-shaped aerodynamic fairing, which smooths the airflow over the rotor hub and reduces turbulence that had generated tail buffeting in the early flight test program. The elastomeric lead-lag dampers on the 500D rotor blades, which replace friction dampers used on earlier models, also are checked as part of the rotor inspection.

Engine starting procedures for the 500D are relatively simple. The throttle is placed in the cutoff position, the ignition key is turned on and the battery/external power switch is placed in the proper position. The generator then is switched on and off to check the audio warning system and all warning lights are pressed to test.

The start fuel pump is turned on and the starter switch at the end of the throttle twist grip is depressed and held. When the gas generator (N_1) turbine speed reaches 12–15%, the throttle grip is rotated through a spring-loaded stop to the ground idle position for engine ignition. Turbine outlet temperature is monitored to ensure engine ignition has occurred and to guard against a hot start. The turbine outlet temperature gauge has a red dot set at 843C to denote the overtemperature limit that can be tolerated for intervals of up to 6 sec. during power changes. During starts, the overtemperature condition can be tolerated for 10 sec., with a momentary peak at 927C, but this limit is not marked on the gauge.

At 58–60% N_1 turbine speed, the starter button is released and the N_1 speed is stabilized between 61% and 65%. With all warning lights out, the throttle is twisted smoothly to the full open position, and the power turbine (N_2) is stabilized at between 103% and 104%.

Takeoff on the first flight from Hughes Airfield here, which has an elevation of 8 ft., required about 57 psi. torque at the reduced gross weight. Outside air temperature was 22C (72F). The helicopter showed a tendency to lift off right skid first, and Hodgson said company pilots have adopted this as standard procedure. The helicopter was stabilized on the left skid momentarily and then lifted into a hover.

Vertical touchdowns also tend to follow the slight left tilt pattern. Once stabilized in a hover, pressure on the collective control is relaxed and the helicopter settles onto the left skid first.

This editor found the 500D to be very responsive in hovering maneuvers. The wind velocity ranged between 10 kt. and 15 kt. during the flight, which provided an opportunity to evaluate the effectiveness of the tail rotor in downwind and crosswind maneuvers. At no time was there any indication that left torque pedal limits were being approached at the reduced gross weight.

Considerable forward cyclic pressure is required to achieve the relatively steep pitch-down attitude used during acceleration onto the recommended takeoff profile. At 30 kt., cyclic pressure is relaxed and the nose pitches up smoothly to the 60-kt. best-rate-of-climb speed.

The standard cyclic trim motor on the 500D takes 11–12 sec. to compensate for the pitch-up and left roll tendency encountered during acceleration through translation lift. This lags behind the changing trim requirement, and this editor found it necessary to hold fairly heavy right cyclic pressure until the trim motor caught up. A high-speed trim motor has been developed for the 500D, and initial flight tests show it has a response time of about 4.4 sec.

Agility of the 500D was demonstrated by Hodgson in a series of high-g maneuvers that included tight turns at bank angles of 70–80 deg., progressive dutch rolls that carried the helicopter through momentary banks approaching 90 deg., wing overs and accelerated quick stops at high pitch angles. Minimal rotor speed degradation was noted throughout the maneuver series.

Purpose of the demonstration was to show the wide performance envelope available with the 500D, Hodgson said. Few customers will use this in normal utility operations, he said, but the high maneuvering limits provide an increased safety margin during emergencies or operations in confined areas.

A series of traffic patterns then was down by this pilot, and the 500D proved to be stable and responsive in all standard approach profiles. Right torque pedal requirement during right turns onto the base and final approach legs was higher than this pilot had experienced on other helicopters.

Landing approach

Autorotation approaches were initiated by chopping the power sufficiently to split the rotor/engine rpm needles, lowering the collective and holding a pitch attitude that produced a 60-kt. descent. Maintaining the throttle above idle keeps the low-rotor rpm warning system active. The throttle then was closed to the ground idle stop and the landing flare was initiated at an altitude of about 30 ft. A countdown to five helped establish the rhythm for pitching forward out of the flare and pulling collective for a running landing or opening the throttle for a power recovery. Minimum rotor speed limit for autorotations is 410 rpm, and the maximum is 533 rpm. Running landings may be made up to 30 kt. on a smooth surface.

Attention has to be paid to the maximum rotor speed limit during autorotations, since an over-speed condition can be encountered with the low-inertia rotor when operating at high gross weights. Rotor speed is easily controlled with collective inputs.

During one autorotation, Hodgson showed how the low-inertia characteristics of the 500D rotor can be used to extend the glide to an emergency landing spot. This is accomplished by pitching over to the 80-kt. best-angle-of-glide speed and raising the collective to slow the rotor to the minimum autorotational speed limit of 410 rpm. Rotor speed is regained by lowering the collective prior to initiating the landing flare.

Autorotations from hovering altitude proved to be gentle maneuvers, due in large part to the low-inertia rotor system. Even when the power was chopped abruptly, little right torque pedal was required to maintain heading. Soft touchdowns were relatively easy to accomplish by holding collective momentarily and then raising it to cushion the impact.

In the high-speed flight regime, the 500D has a maximum autorotational speed of 130 kt. An autorotation could be initiated at a higher forward speed, even though it would involve an excessive rate-of-descent, Hodgson said. But the certification requirement that all limits be exceeded 111% made it impractical to set the maximum autorotation speed higher.

The first flight was completed with a high-speed run out to the beach along the Pacific Ocean just west of Hughes before returning for a final landing. A speed of 155 kias. was reached in a shallow dive, with very little increase in vibration and noise levels. In all flight regimes, including maximum speed at maximum takeoff power, the 500D was considerably smoother than any other turbine-powered helicopter this editor has flown.

On the second flight, the wind was a steady 5 kt. and the outside air temperature was 20C (68F). The first point demonstrated by Hench was the 910-fpm vertical rate-of-climb performance guarantee for a maximum gross weight takeoff at sea level on a standard day. With maximum takeoff power, a vertical climb in excess of 910 fpm was held through 200 ft.

Following a steep approach back to the landing pad, another takeoff was initiated to achieve the 1,680-fpm maximum rate-of-climb point. With maximum takeoff power and a 75-kt. climb profile, a steady rate-of-climb of 1,600 fpm was held to an altitude of 1,100 ft. where the outside air temperature was 18C (64F).

Maximum speed performance was demonstrated during a run northward along the Pacific coast beach at a pressure altitude of 160 ft. (density altitude equivalent to sea level). At maximum continuous power, the indicated airspeed ranged between 141 and 142 kt.—1 to 2 kt. above the guarantee.

Once clear of the terminal control area at Santa Monica Airport, a 1,400–1,500-fpm climb was initiated to altitude for a demonstration of the hover out-of-ground-effect ceiling. A hover was held at a pressure altitude of 7,500 ft., which at an outside air temperature of 6C converted to a density altitude of about 8,200 ft. at maximum continuous power. Even at this altitude, the helicopter retained a slight vertical climb capability.

The helicopter then was accelerated to 65 kias., and a 1,000-fpm climb was initiated toward the 15,200-ft. service ceiling listed for the 3,000-lb. maximum gross weight. By this time, 110 lb. of fuel—slightly over one quarter of the 412-lb. maximum fuel load—had been consumed.

By the time an altitude of 12,000 ft. had been reached, the rate-of-climb had stabilized at 500 fpm at maximum continuous power and at a speed of 50 kias. This was held through 14,000 ft., at which point the climb was terminated, since it was evident the continued 500-fpm rate-of-climb would make it possible to exceed the 15,200-ft. ceiling. Outside air temperature was –6C (21F).

At the peak altitude, a constant watch had to be kept for commercial jet transports descending from the north as they began their approaches to Los Angeles International Airport. A gradual descent, ranging between 500–1,000 fpm was initiated to ease the pressure changes on the pilots' ears and the helicopter was flown back to Hughes for a final landing.

Engine shutdown procedures followed for the 500D include a 2-min. cooling period at ground idle to allow the turbine outlet temperature to stabilize. This eases the thermal cycle during shutdown and reduces engine wear. The spring-loaded idle stop then is lifted out of its detent and the throttle is twisted to cutoff.

Production rate

The company has delivered about 40 D model helicopters to date, according to Thomas R. Stuelpnagel, vice president and general manager of Hughes Helicopters. Tie production rate is expected to stabilize at about 17 machines a month, compared to the rate of 10 a month maintained for the C model.

Certification of the D model was completed late last year, about six months behind schedule, delaying the start of production deliveries. Some problems were encountered with elements in the main and tail rotor transmission, which did not handle the increased power output as expected. Most of these were attributed to quality control difficulties associated with the start-up of new gear production by suppliers, Stuelpnagel said, but the problems were solved without having to redesign the parts involved.

The Federal Aviation Administration also decided to treat the 500D as a new helicopter instead of a new model, so the certification process took considerably longer than planned, he said. Hughes decided to retain the basic 500 airframe for the D model because it combines all the elements the company believes it needs for this specific portion of the market, he said.

The 500 series helicopters can carry two passengers in the front seat, in addition to the pilot, and two in the back seat. But the front seat is crowded with three persons and the head space in the back seat is obstructed by the main drive shaft housing that protrudes through the cabin ceiling.

Stuelpnagel acknowledges that the limited cabin space and the lack of a separate baggage compartment are factors that might limit the 500D's appeal for some executive transport operators. But the performance, versatility and crash safety characteristics of the design apparently outweigh these limitations, because the helicopter has established a place in the world market, he said. Basic price for the 500D is $190,000, which includes some equipment that other light helicopter manufacturers list as optional.

Ease of handling stressed for AS350

Robert R. Ropelewski/Marignane, France
January 17, 1977

Low manufacturing and operating costs and minimal maintenance requirements of Aerospatiale's new AS350 light helicopter will be given heavy emphasis when the French firm intensifies its marketing efforts with the new aircraft in the next several weeks.

Designers of the AS350 in Aerospatiale's helicopter division here believe that the new manufacturing methods and advanced but simplified rotor technology they have incorporated in the five-six place helicopter will reduce the cost per pound of payload flown to about one-third that of the current generation of light turbine helicopters.

Reduction in selling price, relatively high payload, moderate speed, high reliability and ease of maintenance were among the design aims laid down in 1972 when Aerospatiale began looking for a new helicopter model to replace its Alouette 2 and Alouette 3 series. A small team of designers and production engineers was assigned design definition and initial development tasks here in what was essentially a "skunk works" type operation.

First prototype of the AS350, called the Ecureuil (Squirrel), flew its maiden flight in June, 1974, and a second prototype followed in February, 1975. First of eight preproduction aircraft made its maiden

The AS350 is configured to carry two pilots and either three or four passengers in the rear cabin.

flight last November. A second preproduction aircraft was rolled out late last December and will be exhibited and flown at the Helicopter Assn. of America's annual convention in Anaheim, Calf., in early February. An English/American name for the helicopter has not yet been announced, but this will probably be done at the HAA convention.

This *Aviation Week & Space Technology* editor recently test flew the AS350 at Aerospatiale's Marignane production and flight test center, and found it surprisingly conventional to handle despite the significant new technology it employs. The evaluation flight of approximately 1 hr. was flown with four passengers on board and covered most of the routine and emergency maneuvers the aircraft is likely to see in service.

Many of the structural features of the helicopter will have only limited direct impact on the pilots who eventually fly it, but are intended to reduce procurement and operating costs for the operator. Main structural elements have been consciously designed to reduce production man-hour requirements. Right-side and left-side components, for example, have been designed to be interchangeable wherever possible. Most of the structural framework is made of pressed light alloy.

The cabin enclosure is made of thermoformed plastic half shells assembled by ultrasonic welding. Single main fuel tank, located behind the cabin, is of polycarbonate plastic spin-molded in one piece. Landing gear is the metal skid type.

Similar efforts at simplification have been applied to the dynamic components of the AS350. Three-bladed main rotor system is made almost entirely of glass fiber and plastics. The blades themselves are made of laminated plastic around a spun glass fiber spar. The blades are close to the same size and profile as those used on the Aerospatiale five-place SA341 and SA342 Gazelle helicopters. But compared to the hand-made plastic blades that Aerospatiale used on the early Gazelles, the blade manufacturing process has now been mechanized.

Aerospatiale's helicopter division officials say the production price of the plastic blades is below that of the metal blades employed on the company's Alouette 2 and Alouette 3 helicopters. Each blade is molded in a single operation, and the leading edge is then capped with a steel plate to combat sand and dust corrosion. The AS350 blades may be among the cheapest in the world to produce, according to Rene Mouille, Aerospatiale helicopter division chief designer. Additionally, he said, they are fail-safe and have an infinite life.

The patented Star Flex rotor head, for which Mouille is largely responsible, also makes extensive use of composite materials. It behaves like an articulated rotor head, but without conventional articulation

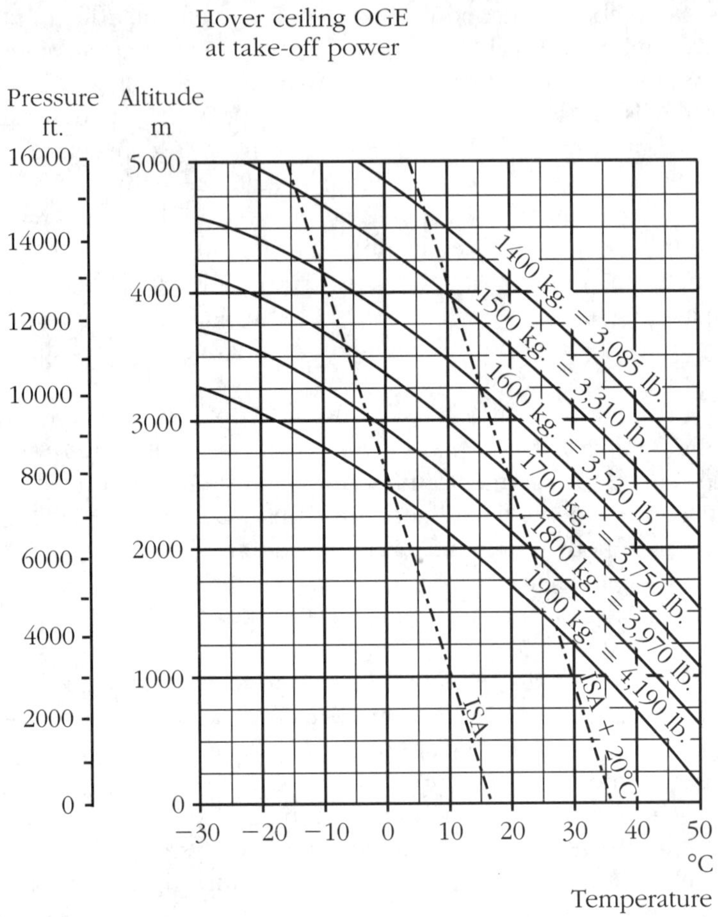

Preliminary performance data for the Aerospatiale AS350 helicopter powered by a single 680-shaft-horsepower Turbomeca Arriel turboshaft engine indicates that at its maximum takeoff weight of 1,900 kg. (4,180 lb.), the AS350 will have a maximum hover ceiling out-of-ground-effect (left) of 3,250 meters (10,660 ft.) and a hover-in-ground-effect ceiling (right) of 3,800 meters (12,464 ft.).

and therefore without the weight and mechanical complexity of a conventional articulated system.

The three-armed hub, thick at the center with the arms tapering off in essentially triangular shapes, is made of laminated glass fiber. The arms of the hub drive the blades in rotation and restrain the blades from drooping too low when the rotor is stopped.

Traditional blade-sleeve attachment to the hub has been abandoned in this case, and the blade, roots are built in the shape of a

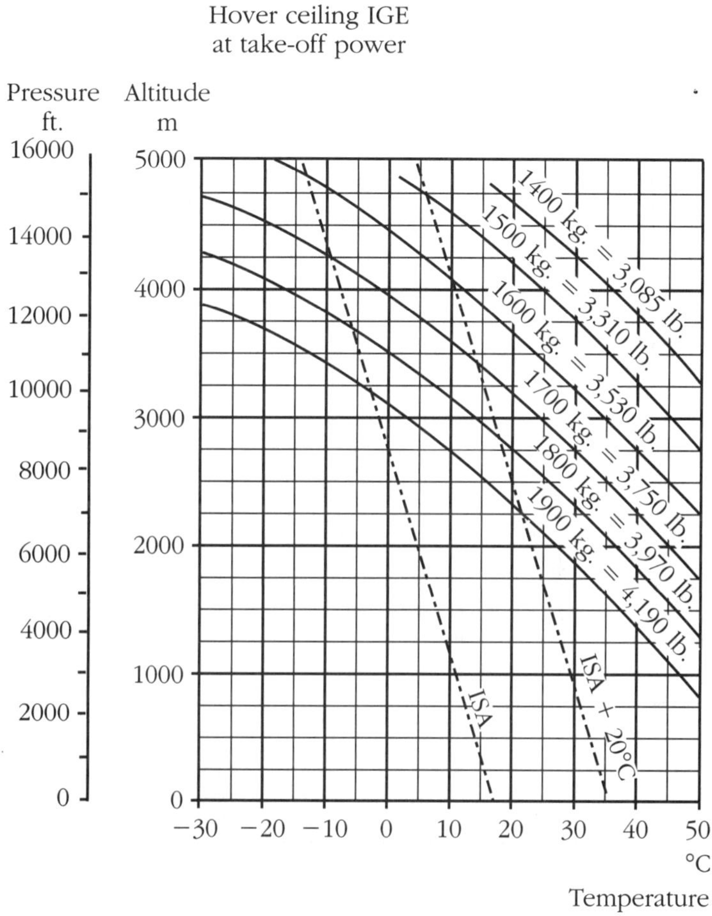

yoke. The yokes are attached not to the arms but to elastomer laminate spherical thrust bearings closer to the center of the hub. The spherical thrust bearings allow blade pitch changes as well as flapping and lead-lag movements, thus eliminating the need for the lubricated bearings used for this purpose on conventional rotor heads. The thrust bearings also absorb blade-generated centrifugal forces when the rotor is turning.

The total number of components in the Star Flex rotor head is only 17% of that of a conventional articulated rotor, according to Aerospatiale. Any part of the head can be inspected visually without removing any other part, and can be replaced in the field in a maximum of 10–15 min. by the operator. Because of the materials and manufacturing processes used, the Star Flex head weighs about half as much as a conventional rotor head of the same size, and production cost of the head is about 40% of the cost of a conventional articulated rotor head.

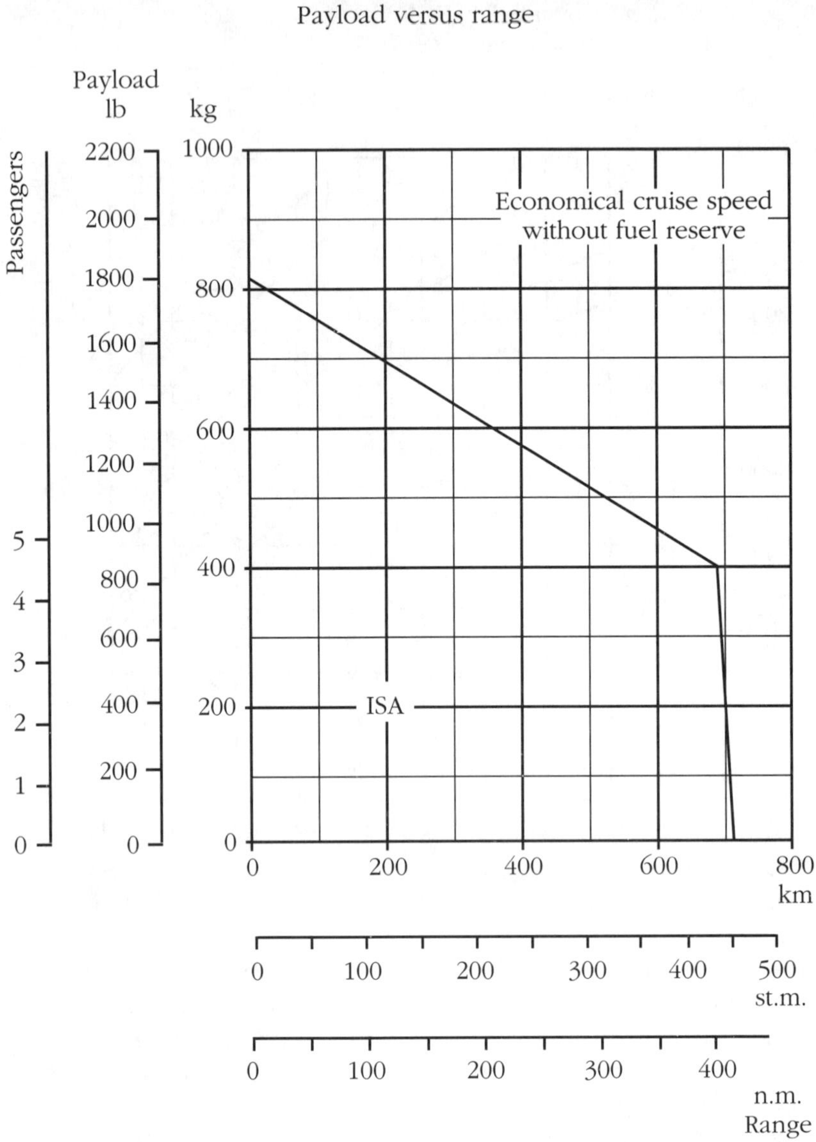

This chart indicates the aircraft will have a range of about 370 naut. mi. without fuel reserves when carrying 5 passengers.

Main gearbox also has been simplified compared to earlier units such as that used on the Alouette 2. It is also oversized for this aircraft, providing large torque margins. The AS350 gearbox has nine main gears compared to 22 on the Alouette 2 gearbox. All but one bevel gear pair have been eliminated on the AS350 gearbox. This has resulted in the adoption of a wider and somewhat heavier transmission housing, but this is offset, according to Aerospatiale engineers, by the fact that the gearbox should never have to be removed during the life of the aircraft.

A simplified two-bladed tail-rotor has been adopted for the AS350 rather than the fenestron ducted-fan/tail-rotor the company employs on its SA341 and SA342 Gazelles. Higher cost of the fenestron was the deciding factor. Helicopter division engineers concede that they still do not have an inexpensive solution to the shrouded tail-rotor concept, though they are continuing to explore this area.

For the AS350, the simplification of the tail-rotor has been accomplished through the use of glass fiber blades on a one-piece roving glass fiber spar and head. Many of the normal mechanical parts associated with a conventional tail-rotor have been eliminated by making the rotor blades flexible in twist for pitch changes, and in the flapping plane to eliminate the need for hinges.

Aerospatiale is developing the AS350 to use two different engines—the 600-shp. Lycoming LTS101 for the North American market, including the U.S., Canada and Mexico, or the 650-shp. Turbomeca Ariel turboshaft engine for France and the rest of the world market. Both are free-turbine designs. Decision must still be made as to whether the Lycoming-powered version of the AS350 will be offered elsewhere in the world if customers request it.

In another effort to keep costs down, Aerospatiale engineers have opted to use low-cost, off-the-shelf components from on the AS350. Among other items, this has included oil radiators and pressure and temperature gauges. Program officials estimate they have been able to reduce costs for some of these items to as little as one-tenth or one-twentieth the prices paid in the past for such equipment. Aerospatiale was obliged to demonstrate the feasibility of this approach to French civil aviation officials.

Apart from the engines, Aerospatiale's helicopter division engineers hope to certify most major drive train components for infinite life. This is not yet the case for the Star Flex rotor hub, but officials hope to have it certificated for the lifetime of the aircraft itself by the time the AS350 is certificated around the middle of this year. For the remainder, the target has been to use components with high mean times before removal.

Because of the overall reduction in number of parts, the attention given to reducing production complexities, and the use of readily available materials, program executives believe the AS350 will require 3.5 times fewer man-hours to produce than either the Alouette or the

Simplified instrument panel of the production AS350 helicopter is patterned after that of an automobile, and allows excellent forward visibility. Circuit breakers, radio selectors and caution/warning lights are on center pedestal. Production version of the AS350 five-to-six-place helicopter will have two hinged doors on each side of cabin, providing easy access for pilots and passengers and for cargo when passenger seats are removed.

Gazelle. The sale price of the AS350 will not be proportionately lower, they caution, because of the need to absorb and amortize general operating expenses within the program budget.

Relatively high sales price of the Gazelle and its subsequent lower competitive position against U.S.-built helicopters in the four-six place civil helicopter market, has been a factor in development of the more-simplified AS350. The AS350 is designed and priced specifically to be competitive against the Bell 206B Jet Ranger and 206L

Long Ranger. Price of the AS350 is expected to be just under $200,000 in 1978, compared to about $180,000 for the Jet Ranger.

A Jet Ranger pilot would not find the AS350's handling characteristics much different from those of the Jet Ranger, but there would be a noticeable difference in performance because of the higher power available with either the Lycoming or the Turbomeca engine on the AS350. This editor found it surprising that the new technology employed in the AS350 did not have more of an effect, one way or another, on the aircraft's handling characteristics compared to other helicopters in the same category.

The *Aviation Week & Space Technology* evaluation flight was flown in the second AS350 prototype, which is powered by the Arriel engine. Right seat was occupied by Aerospatiale test pilot Henri de Riols de Fonclare, who initially demonstrated some of the AS350's capabilities and then served as a safety pilot for the remainder of the flight.

Cabin length and configuration of the second prototype varies somewhat from the final production version, but handling and performance features are essentially the same as those that will be found in the production AS350, according to de Fonclare. This aircraft had logged about 300 hr. of flight tests up to this point.

A walkaround inspection before the flight was routine, and no climbing required to get closer to any items. Top of the rotor mast is just over 9½ ft. above the ground, and with no oil or hydraulic reservoirs to check on the hub, visual inspection could be accomplished from the ground. Transmission oil level and hydraulic reservoirs for the aircraft's three control servos can be viewed either by opening a large access door on the left side of the aircraft, or by peering through a small window in the door.

There is a baggage storage space on both the left and right side of the fuel tank inside the fuselage, and an additional baggage space with separate access door behind the fuel tank in the extreme rear of the main body of the aircraft. Combined volume of the three spaces is about 1.15 cu. meters (35.29 cu. ft.). Opening either of the forward baggage doors provides a visual check of fuel quantity, because the 140-gal. spun-fiber fuel tank is semi-transparent and the fuel level can be seen easily.

Interior of the cabin was comfortable and pleasantly decorated. Instrument panel in this particular aircraft stretched the full width of the cabin because it included flight test instrumentation. Production versions of the AS350 will have a reduced instrument panel that will allow excellent vision for occupants of both front seats.

Production version of the AS350 also will provide more leg-room for passengers than was available in the second prototype. Four tall passengers were carried during this particular flight, and while there was ample knee-room for all of them, there was little remaining space to stretch their legs between the back and the two front seats.

Bench-type rear passenger seat can be configured to accommodate either three or four passengers, giving the aircraft a total seat capacity of either five or six, including the pilot. Customers will be able to procure a kit, consisting mainly of armrests and different cushions, with which they can change from the four passenger to three-passenger rear-seat configuration. Total cabin volume for the production AS350 will be a comfortable 76.28 cu. ft., with access provided to either the pilots' or the passengers' seats by a single large, hinged door on either side.

Outside air temperature was 13.5C when the Arriel engine was started using the on board battery power supply. Start was accomplished by pushing the starter until engine rpm reached 40%, then advancing the single fuel lever to the full forward position. Fuel lever is located on the floor between and slightly forward of the two pilots' seats. Rotor rpm built up slowly as the engine speed wound up, until rotor rpm reached its normal operating range of 3,700–3,900 rpm.

Total weight of the aircraft as it was lifted into a hover was about 1,790 kg. (3,938 lb.) including the six occupants and 280 kg. (616 lb.) of fuel. Normal maximum takeoff weight of the production AS350 will be 1,900 kg. (4,180 lb.), or 2,100 kg. (4,620 lb.) with an external load.

Aerospatiale's helicopter division engineers made an effort to keep cabin noise levels comparatively low on the AS350. This was not immediately apparent when the helicopter was first started, and the cabin noise levels seemed to be about the same as those of other helicopters in this class. However, when power was increased to begin flying, the noise in the cabin stayed at about the same levels, allowing raised-voice conversations throughout most of the flight.

Torque limit

Cool weather and the margin between the actual aircraft weight and the maximum allowable takeoff weight permitted the aircraft to be operated comfortably within the turbine-inlet temperature limits of the engine (740C maximum continuous) and the transmission torque limit (82%). Current torque limit includes a wide safety margin, according to de Fonclare, and this limit is expected to be raised by about 20% as flight testing proceeds.

There was no noticeable vibration signaling translational lift as the aircraft was accelerated from hover into forward flight, and airspeed built up quickly to 70 kt. With the engine at its maximum compressor speed of 99% and the airspeed held between 65 and 70 kt., the vertical speed indicator showed a climb rate of 2,000 fpm until the aircraft was leveled at 3,500 ft.

At a fast cruise of 120 kt. at this altitude, engine rpm registered 96%, torque 75% and turbine inlet temperature 650C. Collective pitch lever friction knob held the pitch steady, and the cyclic stick could be released from 2–3 up to 10–15 sec. before the aircraft started a slow roll in one direction or another. No stability augmentation is provided on the AS350, but controls will be rechanneled on production versions of the aircraft as a provision for possible later installation of an artificial stabilization system.

Vibration levels

Like all of Aerospatiale's recent helicopters, vibration levels were comparatively low, whether in a hover, in level flight, or in 45-deg. banked turns at 120 kt. The three-bladed rotor system, use of elastomeric bearings to suspend the main gearbox from the airframe, and the use of bifilar shock absorbers in the rotor head, all contribute to the low vibration levels at normal speeds. Vibrations increased noticeably when speed was increased from 120 to the maximum allowable speed for the AS350 of 147 kt. These vibrations were most noticeable above 140 kt., and at 147 kt. were at a level where instruments tended to become increasingly difficult to read.

A 2g turn at this point did not noticeably increase the vibrations. Aerospatiale test pilots have so far flown the AS350 up to a speed of 170 kt., and they believe the maximum speed of the aircraft can be safely increased by 3–4 more kt. to around 150 kt. as testing proceeds.

Aerospatiale pilots also have had the aircraft up to 23,000 ft.—about 3,000 ft. higher than the anticipated service ceiling of 20,000 ft. A series of landings at a small grass field at an altitude of about 1,500 ft. showed the AS350 to be easily controlled despite the absence of stability augmentation. The sole problem for an out-of-practice helicopter pilot was the very sensitive rudder pedals, but some practice in the hover smoothed this out.

de Fonclare cut the engine at an altitude of about 600 ft. above the field at one point during these exercises and demonstrated a smoothly executed autorotation down to a firm but comfortable slow-speed run-on landing on the grass field. Airspeed was kept at about 70 kt. until the final portion of the autorotation. This is a normal approach speed for the aircraft.

Aerospatiale's test pilots have taken the AS350 through most of its normal operating envelope in the first 600 hr. of testing with the two prototypes. About the only problem they have uncovered with respect to the Star Flex rotor head has been a tendency of the aircraft to roll sharply to the right in case of a hydraulic failure affecting the control servo for the roll axis.

Redundant servo

One solution to this would have been to incorporate a redundant servo system, but this was rejected because of the added cost and complexity it entailed. Because it is possible for the pilot to counteract the adverse roll force when the aircraft is flying at less than 80 kt., the alternative selected was to install a hydraulic accumulator that compensates for about 1 min. for the right roll forces on the stick in case of hydraulic failure. This is enough time for the pilot to recognize the situation and slow the aircraft to 80 kt. or less, at which speed the aircraft is controllable.

This situation was examined in flight, with de Fonclare turning off the hydraulic boost system while this editor was flying the aircraft at a cruise speed of about 120 kt. Even with the accumulator helping to resist the roll force, the loss of the hydraulic system imparts a strong rolling moment to which the pilot must react quickly to stop the aircraft from entering an unusual attitude.

Once this was under control, power was decreased and the aircraft was slowed to 80 kt. At that speed, the stick pressure to the right remained, but the aircraft could be adequately controlled and could have been flown that way indefinitely, if not somewhat uncomfortably.

It was perhaps the tendency to think of the Star Flex rotor as a rigid rotor system that accounted for some of the surprise when the AS350 was flown. It is not a rigid rotor system, and this was underscored by the control responses. Rotor responses to cyclic inputs had more of the sluggishness of an articulated rotor system rather than the relatively acute sensitivity of a rigid rotor system.

One noteworthy aspect of this flight was the performance of the Arriel engine. No matter how quickly power was called for or reduced during the flight, the free turbine rpm varied little more than 150–200 rpm (about 5% maximum) in either direction. At one point during cruise, collective pitch was moved quickly from minimum to maximum and back again several times without significantly affecting the engine's behavior.

Aerospatiale test pilots on the AS350 program said that the Lycoming LTS101 engine installed on the No. 1 prototype has not shown the same fast response to collective movements, but they added that Lycoming is working on an electronic fuel control system for the engine that will correct this.

Time and weather prevented an examination of the AS350's performance capabilities at high altitudes, but Aerospatiale engineers contend that the aircraft will have significantly better hover-in-ground-effect and hover-out-of-ground-effect capabilities at high altitudes than its predecessors or its current competitors.

Given the weight of the aircraft (about 3,900 lb.) on this particular flight and the ambient atmospheric conditions, performance charts for the AS350 say the aircraft should have been able to hover out of ground effect at an altitude of 10,200 ft., and in ground effect at an altitude of 12,500 ft., using takeoff power in both cases. Preliminary performance data for the AS350 with the Arriel engine also indicated that the aircraft's range in these circumstances should have been about 370 naut. mi. if the helicopter had been flown on a straight point-to-point mission.

All of these figures would be somewhat lower for the Lycoming engine, although Lycoming is now working to increase the takeoff power of the LTS101 engine to 650 shp. by next March, and to 675 shp. by November, which would put the Lycoming-powered AS350 in the same performance bracket as the Turbomeca-powered version of the helicopter. Turbomeca, in the meantime, is continuing its efforts in France to increase further the power of the Arriel.

AS350 program officials expect to use a total of eight preproduction aircraft in the AS350 flight test program, with the last of these

Aerospatiale AS350 was developed to use both the Lycoming LTS101 and the Turbomeca Arriel turboshaft engines.

rolling from the assembly line in November. All of these aircraft will be at or close to production standards as far as their components are concerned, but the actual manufacturing methods may vary somewhat from the definitive production techniques planned. A few changes have been adopted from the prototypes to the production standard aircraft to reduce manufacturing complexity even more.

Certification of the AS350 is anticipated around the middle of this year by both the French Direction General a l'Aviation and the U.S. Federal Aviation Administration. Aerospatiale officials have been meeting with representatives of both agencies to work out the certification program that will be required. Contacts also have been made with German, British and Canadian civil aviation authorities to obtain certification of the AS350 in those countries.

Agusta A-109 offers avionics options

Donald E. Fink/Burbank, Calif.
March 1, 1976

Italy's Agusta A-109, a twin-turbine powered eight-place helicopter, offers corporate/commercial operators relatively smooth high-speed cruise performance, good single-engine safety margins and a variety of optional avionics packages that will enable customers to choose the operational capability best suited to their requirements.

Atlantic Aviation Corp. has exclusive distribution rights for the A-109 in the U.S. and Canada and will have 10 machines available for delivery this year. Agusta's planned A-109 production for 1976 is 100 machines. Atlantic's five-year sales agreement with Agusta covers a total of 186 aircraft, with an additional 34 to be shipped to the U.S. in 1977, 42 in 1978 and 50 each in 1979 and 1980.

The A-109 is certified already in the U.S. and will complete Ministry of Transportation certification in Canada late this year. Deliveries to Canadian customers could begin next year.

Agusta currently has 10 A-109 pre-production helicopters and two production models involved in flight test and flight demonstration programs. One of the pre-production models—N-109W—is being operated in the U.S. by Atlantic, and was evaluated here recently by this *Aviation Week & Space Technology* pilot.

Agusta A-109 twin-turbine, eight-passenger helicopter features a highly streamlined fuselage and retractable tricycle landing gear, which give the aircraft a 150-kt. cruise speed and a 167-kt. maximum speed. Atlantic Aviation Corp. has North American distribution rights on the A-109 and will have 10 machines for delivery this year.

The pre-production A-109 is powered by two 400-shp. Detroit Diesel Allison 250 C-20 turboshaft engines, while 420-shp. C-20B engines are used in production models. Other changes in the production configuration include elimination of a nonstructural partial bulkhead that formed the back of the pilot and copilot seats. Individual tracked seats are now used for the two forward positions, adding another 3 in. of leg room to the passenger compartment.

Evaluation flight by AW&ST lasted about 30 min. and included two takeoffs and landings from the Hollywood Burbank Airport. Sanford Kaplan, chief pilot for Atlantic Aviation's Agusta Div., was in the right seat.

The helicopter had a gross takeoff weight computed at 4,640 lb., including an airframe configuration weight of 3,400 lb., 480 lb. of fuel, two 180-lb. occupants and two who weighed 200 lb. Normal operating gross weight for the A-109 is 5,400 lb.

Engine start was accomplished using onboard battery power, with engine light occurring at 15% rpm. The starter/generator was then brought on line to recover the battery prior to starting the other engine.

The 28-volt dc electrical system in the A-109 is energized by dual 150-amp. engine-driven starter/generators. A 13 amp./hr. nickel-cadmium battery is standard equipment, but a 22 amp./hr. battery is offered as an option.

A noticeable lateral vibration developed just prior to liftoff, but disappeared when a load was applied to the rotor blades. This was caused by a defective blade damper that was due to be replaced.

A rolling takeoff was made, and the A-109 accelerated rapidly through the translational lift phase. The rate-of-climb stabilized at about 2,000 fpm after gear retraction. Gear retraction was smooth and soundless.

The A-109 has a tricycle landing gear with a castering nosewheel, which can be locked for takeoffs and landings. But the self-centering spring mechanism, which disengages when the gear is retracted, is sufficient to hold the nosewheel for most ground operations, Kaplan said. Taxi tests have been run up to 60 kt. with the nosewheel unlocked without encountering shimmy problems, he said.

Noise and vibration levels in the cabin were relatively low during the climbout, due partially to the A-109's four-bladed semi-articulated rotor and good cabin soundproofing. Peak vibration level was reached at about 135 kt. and stayed constant throughout the acceleration to the maximum sea level indicated airspeed (V_{ne}) of 167 kt.

Kaplan said that during early evaluation tests with the A-109 he had reached a maximum speed of 205 kt. in a long shallow dive. Vibration levels had increased somewhat, but the machine responded normally and showed no evidence of encountering retreating blade stall, he said.

Single-engine performance

Single-engine performance was demonstrated several minutes into the flight by shutting down the left engine. Airspeed in level flight stabilized at 110 kt. The nose was raised and the single-engine rate of climb ranged between 600 and 700 fpm, at an indicated airspeed of 65 kt. The single-engine climb was started at 2,500 ft. with an outside air temperature of 10C.

Most noticeable handling characteristic of the A-109 is its lateral sensitivity. This pilot and another experienced helicopter pilot participating in the flight demonstration both showed a tendency to overcontrol the machine in the lateral axis. Attempts to anticipate and correct lateral movements often resulted in pilot-induced oscillations.

Kaplan said the lateral sensitivity had bothered him for about the first hour during his transition training in the A-109. Once he became accustomed to the characteristic it posed no problem, he said.

Control response good

The A-109 showed good control response in turns of varying bank angles and exhibited basic handling qualities that gave a sense of solid feel. Hands-off stability was good for a machine without stability augmentation.

Agusta is offering individual stability augmentation systems (SAS) links for any of the three axes as well as a three-axis SAS unit in its optional equipment packages, Kaplan said. Some operators who only want a visual flight rule (VRF) capability may elect to add SAS links in one or more axes to reduce pilot workload and ensure a smoother ride for corporate passengers.

Limited visibility in the basin surrounding Hollywood Burbank Airport precluded full hover demonstrations with the A-109 during the flight. Agusta lists hover ceiling in ground effect as 15,000 ft. at a gross weight of 4,100 lb. and 9,800 ft. at the 5,400-lb. normal operating gross weight. Hover ceiling out of ground effect is 12,400 ft. at 4,100-lb. and drops to 6,700 ft. at the 5,400-lb. gross weight.

Single-engine hover out of ground effect has been demonstrated at a density altitude of 3,500 ft. and close to maximum gross weight, Kaplan said. The A-109's ability to make an emergency single-engine takeoff also has been demonstrated, he said. The procedure involves a rolling takeoff to an interim level speed of 40 kt., followed by acceleration to 60–70 kt., at which point a climb of 600–700 fpm can be initiated.

Comfort levels of both the flight crew and passenger seats are excellent in the A-109. Forward and lateral visibility for the pilots is enhanced by a large split-panel windshield that curves around on each side of the forward fuselage. Large overhead windows are placed on either side of the ceiling-mounted engine control quadrant and floor windows are located ahead of the torque pedals for downward visibility during landing flares or hover operations.

A run-on landing was made on the first return to Hollywood-Burbank. The landing gear was extended as the speed dropped below 100 kt. Initial approach was flown at 70 kt. and final approach at 60 kt. Touchdown was made in a slight sideslip, but the main gear units absorbed the side load smoothly and nosewheel castered to straighten out the short landing roll.

The evaluation was concluded with a vertical takeoff to a brief hover prior to acceleration to forward flight, another pattern around the Hollywood Burbank area and a return to a hover and vertical landing.

Following taxi to the parking ramp, the engines were run at 60% rpm for a 2-min. cooling period prior to shutdown. The rotor brake was engaged after engine shutdown and it brought the rotor to a smooth stop in about 6 sec.

First production model of the A-109 will be delivered to Atlantic next month, and the company will use it as its prime sales demonstration machine in the U.S. Aircraft N-109W will go into overhaul to be fitted with C-20B engines, upgrading it to production performance standards.

In the A-109, the C-20B engines are certified for a maximum takeoff power of 420 shp. for 5 min. and a maximum continuous power of 400 shp. The transmission has a twin-engine continuous rating of 692 shp., so a maximum of 346 shp.—86.5%—is pulled from each engine during normal operations. For single-engine operations the engine can be run at 420 shp., depending on the outside air temperature, but its normal certification is 400 shp. for 5 min. and 385 shp. for maximum continuous power.

Wide range of technology

Agusta has applied a range of technology developed through license agreements with major U.S. helicopter companies to the A-109. The company currently is building the full range of Ben Helicopter Textron machines, up to the 15-place Agusta-Bell 212, the Sikorsky SH-3 and HH-3 series aircraft, and the Boeing Vertol CH-47C Chinook. Agusta's association with Bell started in 1952.

With this technology base, the company set out to develop a new twin-turbine design suited to the worldwide corporate/commercial operator market. Wherever possible, proved systems and equipment—much of it from U.S. suppliers—were applied to the design, with the result that import duty on the helicopter is paid on the basis that it has 80% U.S. content.

Bonded honeycomb

The highly streamlined bonded honeycomb fuselage embodies several features aimed at enhancing the A-109's high-speed performance, including the retractable landing gear and flush-mounted, low-drag engine inlets. The fuselage is 35.2 ft. long (overall length is 42.7 ft. with the rotor turning) and the fuselage width is 9.7 ft. Height at the tip of the vertical stabilizer is 10.8 ft., and the rotor diameter is 36.1 ft.

The cabin is slightly over 9 ft. long and 4.3 ft. wide, which allows for a variety of seating arrangements. Standard interior consists of the two crew seats and two forward-facing triple seat benches in the aft cabin.

The aft bench seat is fixed, but custom interiors are available with aft-facing seats in the front part of the cabin and a bar or work table between, according to Richard F. Powers, general manager of Atlantic Aviation.

List price of the A-109 in the standard configuration is $695,000, and optional dual controls and flight instruments add about $3,300. Another $30–$32,000 worth of optional equipment is available, including $1,500 custom interiors.

A variety of optional navigation/communication equipment also is offered, giving the customer the choice of equipping the machine according to his needs. Basic Navcom packages offered by Agusta are the Pro-Line or Micro-Line systems built by Rockwell International's Collins Radio Group. King Radio Corp. equipment also is offered as an option.

This philosophy of offering the customer a range of choices is being followed in the instrument flight rule (IFR) packages being designed for the A-109, Powers said. Agusta and Atlantic are working with Sperry Rand Corp's Sperry Flight Systems Div. to develop the autopilot, flight director and stability augmentation system (SAS) installation to be used for the A-109 IFR certification.

Most of the IFR equipment has been installed in a pre-production machine in Italy, where preliminary engineering work is under way leading up the Federal Aviation Administration certification program.

The IFR machine will be shipped to the U.S. next August and the certification trials are scheduled to start in FAA's Southwest District in November, Powers said. A flight evaluation program of 45–50 hr., with 5 hr. of actual IFR operations, is expected to be concluded by the end of the year.

Atlantic's aim is to certify the helicopter with the minimum IFR package acceptable to the FAA, Powers said. Using the minimum IFR package will give A-109 customers more latitude in selecting the IFR capability most closely suited to their requirements. For example, the certification trials will not be based on operations with a fully coupled flight director.

Using the minimum/maximum approach will give operators who want it the option of ordering a full avionics package that will make the A-109 a single-pilot IFR machine, he said. But other customers who do not need that capability can order a partial package that will adapt the helicopter for dual-pilot IFR operations.

Atlantic, which had been a Bell Helicopter distributor until last summer, is confident the A-109 fills a distinct need in the U.S. and Canadian market, Powers said. The price and performance of the machine are in its favor, as is the fact that it is entering the market two years ahead of its strongest competitors, the Bell 222 and the Sikorsky S-76, he said.

Some European helicopters distributed in the U.S. have encountered spares and support service problems, Powers said, and Atlantic is making a big effort to ensure this does not happen with the A-109.

Spares for 10 more machines than currently are operating in the U.S./Canadian market will be stocked in Atlantic's Wilmington, Del., headquarters, he said. An air freight link also has been established with the Agusta factory to guarantee delivery of any part within 24 hr.

Service centers will be established as A-109s go out into the field, and the intent is to locate some kind of support facility within 2 flight hr. of each machine, Powers said. District offices to support the A-109 also are being established in Teterboro, N.J.; Oklahoma City, Okla.; Chicago and Los Angeles. Atlantic's Houston facility also will be a major service center supporting the helicopter. Atlantic has sent pilots and mechanics to Agusta for training on the A-109 and will offer both flight instruction and maintenance training on the helicopter in Wilmington.

The A-109 is designed for low maintenance operations and easy service, according to Luciano Miotello, an Agusta technical representative working with Atlantic Aviation. The main rotor has a 3,600-hr. fatigue life and the tail rotor a 1,000-hr. fatigue life. All bearings are lined with Teflon and all gearboxes have chip detectors.

An access door aft of the right main cabin door has been redesigned to open downward and serve as an integral step for service crews, he said. This will enable ground crewmen to reach the rotor hub without a ladder or maintenance stand.

Magnetic pickups and interrupters are built into the rotor hub for dynamic balancing of the rotor system, to eliminate time-consuming installation of test equipment, Miotello said. Connectors for attaching strobe light equipment for balance and tracking tests are located in the upper cockpit bulkhead behind the copilot's seat.

All fuel, oil and hydraulic fluid receptacles on the A-109 have scuppers under them to catch spills during filling operations, he said, and provision has been made for installation of whatever radio equipment customers may order by including five VHF/FM antenna attachment points and cables in the aft fuselage and tail rotor boom. A full complement of ducts and vents has been included in the cabin overhead.

A-109 PRELIMINARY PERFORMANCE SUMMARY

Standard Day, Allison 250 C-20B

Mode	**4,100 lb.**	**5,400 lb.**
Hovering I.G.E.	15,000 ft.	9,800 ft.
Hovering O.G.E.	12,400 ft.	6,700 ft.
Certificated ceiling	15,000 ft.	15,000 ft.
Max. speed, sea level (V_{ne})	167 kt./ 192 mph.	167 kt./ 192 mph.
Normal cruise, sea level	150 kt./ 173 mph.	150 kt./ 173 mph.
Sea level max. rate of climb both engines at 60 kt., ISA	2,420 ft./min.	1,620 ft./ min.
Sea level max. rate of climb one engine at 60 kt., ISA	700 ft./min.	300 ft./min.
Max. endurance at 56 gal./hr.		2.6 hr.
Max. endurance speed		78 mph.
Max. range at 56 gal./hr.		385 mi.
Max. range speed		148 mph.

Typical missions	**#1**	**#2**
Cruise: 150 kts./173 mph., sl., ISA standard configuration weight	3,224	3,224
Fuel: (#1: 146 gal.) (#2: 86.3 gal.)	976	578

Typical missions	#1	#2
Loading: pilot + 5 passengers	1,200	
pilot + 7 passengers		1,600
Total weight	5,400	5,400
Distance	451 mi.	267 mi.
Time en route	2.6 hr.	1.5 hr.

Allison 250 C-20B Engine

Takeoff horsepower	420 shp.
Max. continuous	400 shp.
Specified sfc. at max. rated power	648/gal.

Transmission ratings

Twin-engine continuous	692 shp.
Single-engine continuous	385 shp.
Single-engine (5 min.)	400 shp.
Fuel	Jet A, A1 and B, JP-4 & JP-5
Fuel capacity	943 lb. (146 gal.)

Agusta A-109 Weights

*FAA empty weight	2,998 lb.
Basic configuration weitght (empty + oil)	3,024 lb.
**Standard configuration	3,224 lb.
FAA normal gross weight	5,400 lb.
Max. useful load (gross wt.,-FAA empty wt.)	2,402 lb.

* Does not include interior.

** Inclues the standard interior (eight seats), carpet, upholstery and headliners.

Index

Illustration page numbers are **boldface**

B

C